# Linking Mathematics and Language:

Practical Classroom Activities

RICHARD McCALLUM

ROBERT WHITLOW

Pippin Publishing Limited

Copyright © 1994 by Pippin Publishing Limited
380 Esna Park Drive
Markham, Ontario
L3R 1H5

Designed by John Zehethofer
Edited by Sylvia Hill
Typeset by Jay Tee Graphics Ltd.
Printed and bound by Kromar Printing Ltd.


## Canadian Cataloguing in Publication Data

McCallum, Richard D.
   Linking mathematics and language : practical
classroom activities

(The Pippin teacher's library ; 12)
Includes bibliographical references.
ISBN 0-88751-038-8

1. Mathematics – Study and teaching (Elementary).
2. Reading (Elementary). 3. English language –
Study and teaching (Elementary). I. Whitlow,
Robert F. II. Title. III. Series.

QA135.5.M23 1993      372.7      C93-093936-0

ISBN 0-88751-038-8

10  9  8  7  6  5  4  3  2  1

# CONTENTS

## Number

## Probability and Statistics

# INTRODUCTION

The news about the state of mathematics education seems to be moving from bad to worse. The popular press reports that our kids can't solve simple word problems using whole numbers, know very little about algebra, and on the whole are not ready to attack the concepts of higher mathematics. Even more disappointing is the fact that many children would rather go to the dentist than study mathematics. Why *can't* our children do it? Are they just lazy and stupid? Or are our teachers shirking their responsibilities? Can't anyone teach anymore?

Regardless of the perspective presented in the popular press, we firmly believe that every child can learn even the most complex mathematical concepts, and that this competence can develop naturally given time and appropriate learning environments. Learning mathematics does not have to be as torturous an experience as having a root canal. It can — and should — be fun!

Our solution to the crisis in mathematics achievement rests with several common sense notions about how kids learn and about how teachers should structure classrooms to facilitate learning. As whole language teachers we have seen how children's literacy development can be fostered in a natural yet deliberate way, and that children can and do develop a love for reading and writing when literacy is an integral part of their day-to-day lives. We see no reason why mathematics cannot be approached in the same way. In fact, we have witnessed it happening.

The activities presented in this text were designed to foster mathematical competence in the same way that whole language teachers facilitate the development of students' reading and writing abilities. The content may be different when dealing with mathematics instruction but, as you will see, the basic principles of learning involved and the suggested teaching methods are very much the same.

. . . . . . . . . . . . . .

# THE PURPOSE

# OF THIS TEXT

This text provides a set of introductory activities in mathematics for children ranging in age from five to twelve using holistic instruction as a tool for the development of mathematical concepts. For example, when a teacher is beginning a unit in geometry, the geometry activities in this text should precede traditional instruction. Using these activities supplements and enriches existing curriculum.

We hope that once teachers see how much fun mathematics can be they'll take every opportunity to integrate it into language instruction, and to design their own mathematics activities.

*Note:* Throughout this text we abbreviate the word *mathematics* to *math*, which is our common usage. We are aware of the variant *maths*, and hope that readers used to this term are not constantly jolted by our variant.

. . . . . . . . . . . . .

# IDENTIFYING PATTERNS:
# THE KEY TO MATHEMATICS

The activities presented here are designed around several basic assumptions about how math is woven into our day-to-day lives. We believe math to be a logical system of thought that is primarily a tool for solving problems and understanding the world around us. Stated simply:

> [W]hat humans do with the language of mathematics is to describe patterns. Mathematics is an exploratory science that seeks to understand every kind of pattern — patterns that occur in nature, patterns invented by the human mind, and even patterns created by other patterns. To grow mathematically, children must be exposed to a rich variety of patterns appropriate to their own lives through which they can see variety, regularity, and interconnections.[1]

Finding, describing, and understanding patterns is central not only to math, but to the human experience itself. It is through this process of identifying interconnections and regularity that we make sense and meaning of our lives. Our understanding of a natural occurrence, such as the weather we experience, is based upon the identification of the patterns we observe and then put into some meaningful and useful form. The same is true of other life occurrences such as human relationships, reading, and mathematics.

Our approach to teaching the patterns of mathematics is largely based upon our understanding of current language instruction in reading and writing. Identifying patterns in mathematics has much in common with the way in which children come to understand the patterns in language use. We believe that teachers should view the development of math ability in the same way that we view language development. In short, we are asking math teachers to become whole-language math teachers. Insights from language development have led to the development of a "naturalistic learning model" from which current language instruction has developed.

The next section outlines this model of learning and provides the basis for our approach, applying this model to math education.

*Patterns in Oral Language: A Naturalistic Model of Learning*

Language ability develops very quickly in children. It almost seems like magic. By the time they are five years old they have mastered most of the patterns that govern their native language. They achieve this without any formal, school-like instruction: no worksheets, no homework, and no explicit skills training! It's a good thing that children have the ability to construct the basic rules of language because, as David Pimm suggests: "If children had to be taught to speak using the methods by which the majority are taught to read and write, many might not say anything!"[2]

Research shows us that children are active participants in language learning.[3] They identify or "construct" the patterns of language over time as they talk with those around them and attempt to use language in the world. We know that *oral language ability develops within the framework of six principles*. These principles are the foundation of all learning.

**1.** Language Develops in a Social Context.

Children learn language from others, usually adults, who guide them, prompt them, and nurture them. It is through social interaction that children learn language, and through language that the values and beliefs of their culture are transmitted. Children's exposure to language through interaction with others provides them with the opportunities to identify and construct the patterns that govern language.

**2.** Language Development Has an Affective Base.

No one can deny that a strong emotional bond exists between parents and their children — a bond that is reflected in the encouragement that children receive as they learn the rules of language. When toddlers gurgle and goosh in their first attempts at language, we accept all their responses and praise them! In this way we foster their attempts at using language. The support and feedback that children receive as they learn language helps to create a strong motivation — a motivation that encourages children to take risks and make mistakes. Mistakes are a very necessary part of growth and learning: through mistakes and subsequent corrections, children learn about language.

**3.** Language Develops in Functional and Meaningful Settings.

Language is a tool. It gets things done in the world, and an awareness of this quickly develops in children. How long does it take for kids to learn to argue for what they want? Staying up past their bedtime is such a powerful motivator that children will use any and all linguistic devices at their disposal — they even generate new arguments — to make their case. Language development is facilitated when children can see that language is a useful tool in their daily lives.

**4.** Language Growth Is a Developmental Process.

Language ability develops over time and in conjunction with physical maturation and emotional development. Even though kids at five years of age know a great deal about language, they will be fine-tuning and adding to that knowledge all the way through their teen years. Parents are intimately aware that children develop at different rates: some quite slowly and deliberately, while others move forward in leaps and bounds. It is a simple fact of life that no two kids are exactly alike. Thank goodness!

**5.** Language Development Is an Active Process.

Just as it is said that children learn to read by reading, it also holds that they learn language by using language. As anyone who has traveled widely knows, learning a foreign language in a classroom can barely match the learning that occurs when you actually try using the language to get things done in a foreign country.

**6.** Language Development Requires Repeated Exposure and Opportunities for Practice.

As children are given opportunities to practice and experiment with language forms and usage, their abilities increase. Practice is an important ingredient in the development of language ability.

Language ability develops naturally within the constraints provided by these six principles. Children actively construct the patterns of language through repeated exposure to language use in interaction with those around them. This constructive process involves the creation of broader and broader generalizations about the patterns of language.

For example, when young children use the term "goed" to represent the past tense form of "to go," they are providing us with a window into their process of constructing the rules governing the correct form for expressing events in the past tense. In the case of "goed" they have over-generalized a rule that states: "When forming the past tense of a verb, add an -ed at the end." Children are not explicitly taught the patterns governing language, rather they naturally construct them through repeated exposure to the way language is used in day-to-day situations.

SUMMARY

The six principles of learning presented here provide the basis for what many language teachers have described as a "naturalistic" model of learning. When these six principles are present, language development is facilitated. When these principles are absent or thwarted in some way, problems and irregularities develop.

APPLYING THESE PRINCIPLES TO MATHEMATICS

The power of the basic principles discussed above is the fact that they hold true for the development of math ability as well as for oral language. It is our contention that *the basic principles of learning that hold true in oral language development should guide the curricular and instructional decisions teachers make in mathematics instruction* — or, for that matter, any area of the curriculum.

However, solving the problems associated with math achievement is not as simple as attending to these six principles, even though they are very important. Language ability develops gradually over time. As Piaget and Inhelder[4] [5] [6] have shown, children mature as they progress through "stages." These stages are characterized by changes in how they approach thinking and learning. Before applying the practical activities presented in this text, it is important to understand the basic stages that characterize the development of both math and language.

## Basic Stages of Math and Language Development

STAGES IN MATH DEVELOPMENT

Marilyn Burns[7] has outlined three basic levels or stages in the development of mathematical concepts: experential level,* connecting level, and symbolic level. Each level is characterized by an increase in students' ability to think abstractly. That is, as children grow, they develop the ability to deal with concepts and ideas not directly tied to their experience. (Note that this is true for the formation of any new concept, regardless of age.)

## Experiential Level

At the first level, childrens' learning about concepts is based upon physical contact with their surroundings. Knowledge is derived through direct experience with the world. Knowledge of the concept of number, for example, is intially derived from direct activity and manipulation in the real world via counting and sorting. Students at this stage need direct hands-on experience to become familiar with the concept of number. Such manipulation is the basis from which all knowledge is built, regardless of age or developmental level.

## Connecting Level

At the second level, direct experience with a concept is connected with language and other symbolic forms. Connections are made at this level between symbols devised by humans, such as pictures and numbers, and the real world objects of beans, bicycles, toys, etc. Kids at this stage may still need to refer back to concrete objects, but they are beginning to move, over time, toward use of numbers or other symbols to represent objects in the world.

## Symbolic Level

At the third level, symbolic forms such as numbers are freed from activity or manipulation of the environment. At this level children need not always refer back to a concrete object to understand the concept, rather they have absorbed an understanding of the concept and can think about and utilize that concept at any time — with or without reference to the object.

The power and beauty of symbol use is that it allows us to deal with larger and more abstract concepts than would be possible if we had to constantly rely upon real world objects for

---

* She uses the term *concept level* for this stage. We have opted to use *experiential* to avoid possible confusion with other terminology. Despite the change in terminology, the meaning remains the same.

understanding. Yet at the foundation of abstraction lies the real world. Learners must start with something they can touch.

Oral language development follows a similar developmental sequence: it begins with concrete forms and then moves to symbolic forms. At the *experiential*, or pre-linguistic level, children are exposed to events, objects, and actions in the world through direct interaction with the environment. Knowledge of the concept "chocolate," for example, is derived from direct experiences with the taste, sight, smell, and feel of chocolate. Children at this stage are very much tied to the sensory dimensions of experience.

At the next level, *connecting*, direct experience with a concept is *connected* with the forms of oral language. At this stage children come to understand that words, that is, sounds, represent objects and events in the world and that language conveys meaning and serves a function and purpose in the world. When a child is asked after dessert, "Did you like the chocolate?" the questioner is not only asking an informational question, but also modeling the connection between language and activity in the world. At the connecting level, children begin to develop an understanding of the relationship between the physical world and the symbolic forms and functions of language.

At the third level, the *symbolic* level, children operate with symbolic forms. As in mathematics, the symbolic forms of oral language (words) are used as a shorthand for the representation of events and objects in the world. Children at this level need no assistance, for example, identifying chocolate on the menu, or understanding what is meant by the sentence: "Would you like a piece of chocolate?"

In school, language is the main tool for learning the concepts for all areas of the curriculum — and math is no exception. It seems simple, but the idea cannot be overstated: It is through language that children come to manipulate and understand ideas. The more we talk about, write about, and interact with math in the context of the key principles of learning and development, the closer we will be to achieving our goals for math education.

## The Situation in Traditional Reading and Math Instruction

The problems currently facing mathematics instruction also plague reading and literacy instruction and other curriculum areas. These problems, we and others feel, stem from a traditional curriculum that does not attend to the natural learning process.

A brief examination of the current situation in reading and math will help us focus our attention on the types of instructional practices that will help assure the success of our curricular objectives.

### THE SITUATION IN READING

The movement toward the natural learning process of language instruction in reading has helped educators and researchers realize that an inconsistency exists in the process of literacy development. Rather than being a natural process of development — from oral language to written language — kids often have great difficulty learning to read and write. Why is it that children develop oral language facility so naturally, without homework, skill sheets, and the like, yet have such difficulty making the transition to reading and writing? There is no doubt that the processes are different, but nonetheless, they are still language processes. Why does this transition cause children such difficulties?

The answer can be found in the nature of instruction provided in reading and writing. In fact, the dominant practices in reading and writing act to undermine achievement. Many of our instructional practices ignore or openly thwart the basic principles of learning.

The problem with traditional instruction in reading is not with the specific rules or skills that are taught, for there is a general consensus among linguists and educators that these are the patterns involved. Rather, the problem rests with the nature of the instruction employed to teach the rules and skills, and the view of the reading process fostered by the adoption of these methods.

What exactly is the nature of the instructional methods employed in traditional reading instruction? The idea is simple and, on a common-sense level, very appealing: identify the rules and skills involved and then explicitly teach each of them. The scope and sequence charts that accompany many language arts programs represent an overview of the abilities considered necessary in reading, writing, and language.

There are serious flaws to codifying skills and explicitly teaching

9

them. The approach is teacher-centered and short-circuits students' attempts to make sense of the patterns in language. The assumption is that students follow the developmental sequence predetermined by the teacher, and that such instruction facilitates the development of the skills necessary to comprehend written language. But, as we have discussed, no two students are exactly alike — they differ greatly in their experiential backgrounds, their ability to understand and construct how language works, and in their patterns and timelines of development. Traditional instructional methods may not allow students the time necessary to construct the rules and develop the skills necessary, because as is often heard in the teachers' lounge: "I can't slow down, I have to get to vowel digraphs by the end of the week!"

Another flaw rests with the nature of the instruction. As with oral language competence, the natural development of reading "skills" requires that students be given the opportunity to construct the rules and conventions that operate through repeated exposure and practice. Traditionally, such practice is accomplished through the completion of worksheets and handouts, each of which focuses exclusively on a specific skill. Such skill work is often decontextualized, that is, the skill work is removed from functional and meaningful settings. Students may be able to fill in the blank on the worksheet with the correct consonant sound, but they may not know how or where to apply the skill. Students may not be able to transfer this knowledge to a similar "real world" situation.

More importantly, continued exposure to such materials leaves students with the impression that reading takes place not with books but rather with workbooks. Not only is the skill work decontextualized, it may also be boring and tedious, undermining students' motivation to read.

What is the legacy of teacher-centered instruction based on skills and rules? In the United States at least, local, state, and national data consistently show that we have produced a generation or more of readers who don't understand the functions and purposes of reading; who can sound out words but who fail to comprehend (in Jerome Harste's memorable term, kids have been taught to "phonicate," not to read); and who can read, but don't. We have developed readers whose motivation is negligible, if there at all.

But why has traditional instruction produced such a situation? First, using Marilyn Burns' terms, traditional instruction in reading begins on the symbolic level, often before students have adequate experience with the real world and oral language. In initial instruction, the reading process is broken into its constituent parts, letters and sounds, and each part of the process is addressed separately. Rather than simplifying the task and facilitating students' construction of the rules involved, such an approach makes the process much more difficult and frustrating. Beginning instruction in reading at the symbolic level puts the cart before the horse. We expect students to make the leap to operating with purely symbolic forms without the appropriate foundation of experience and language ability. Adopting such an approach guarantees, sadly, that many students fail.

Consider the symbolic level of the introduction of the letter ''b.'' Students may be faced with an obscure abstraction — not with a concept that they have experienced in the real world of their day-to-day lives. Consider what is involved in understanding the letter ''b'' in isolation. First, students must be able to employ what is known as phonemic segmentation — they must be able to conceptualize a word such as *bike* as a series of discrete sounds. Further, they need to know that each of the sounds is associated with certain letters, and in the case of the letter ''b'' that this sound is regular and consistent. This is no small task. If you have ever tried to understand a conversation in a language foreign to you, you know the difficulty: you can't tell where one word ends and the other begins. Imagine trying to read Arabic: not only are the sounds unfamiliar, but the letters make no sense as well.

In traditional, skill-based approaches the focus of instruction is not on the whole (meaning) but on the part (skills). Kids may score full marks on the CVC pattern on their worksheets, but that does not necessarily guarantee they will be able to apply the skill in other situations, or be able to comprehend the basic meaning of a selection, or indeed care greatly one way or another about why they are doing so. Instruction that ignores meaning and reading as a functional language process subverts the natural pattern of language development. It compromises and complicates the learning process.

The same basic situation outlined in reading instruction applies to mathematics instruction in the elementary grades, and with the same sad outcomes. Traditional instruction focuses on skills. We were taught mathematics, and we are still teaching mathematics, via rules: rules for adding, subtracting, multiplying, dividing, and rules for the application of formulae. These rules are learned through rote memorization and repetitive practice of basic ''math facts.''

As with reading, the dominant approach is that of drilling on small often unrelated facts and teacher-centered instruction. As is the case in reading, adopting such an instructional methodology violates the principles of learning.

First, initial instruction in mathematics, like reading, often puts the cart before the horse. Rapid-paced math instruction often begins at the symbolic level with the introduction of numbers, and symbols that represent concepts, without allowing students the time to develop the experiential base necessary to understand these concepts. Numbers, though, are abstractions, and as was the case with the letter ''b'' discussed earlier, students may not have had the experiences that allows them to understand the concepts behind the symbols.

Second, direct instruction in mathematics focusing solely on skill development shifts the focus from the whole (the relationships among math concepts and their application in the world) to the parts (the skills necessary to solve problems). As with reading, students may be able to do math facts worksheets at a high level of accuracy, but not be able to apply that knowledge in other real world situations. Students who cannot understand the larger function and meaning involved in mathematics cannot apply their knowledge to new, real situations.

Third, as in reading, such skill development often occurs in teacher-centered environments. Instruction is driven not by the students' developmental level, but rather by a pre-set scope and sequence of skills developed for the ''average'' student. Further, given the focus on rote memorization and repetitive drill, such instruction may not be sensitive to student's interest and motivation. We turn kids off math.

## Changing Instructional Practices: What Can We Do?

Let there be no doubt that skills play a critical role in success in both reading and mathematics. Children must have these tools when reading, calculating, and solving problems. It is not a matter of should we provide such instruction, but how and when and at what cost.

What we must do now is to reverse the traditional practices that undermine naturalistic learning. We must bring instructional practices in mathematics and reading in line with the principles of learning that are characteristic of language development. This does not mean abandoning a concern for the development of skills, but rather that we reconceptualize how we go about developing competency in mathematics.

The current movement to naturalistic learning instruction in both math and the language arts is an attempt to reverse the situation that exists in these content areas. A discussion follows of some of the changes in instructional practice that have found their way into elementary classrooms. This discussion clarifies the preceding discussion and provides a foundation for the practical activities in this text.

HOLISTIC INSTRUCTION IN READING

Although the movement to reading instruction sensitive to the six principles of learning has gained momentum and many converts in the last few years, instructional approaches based on these principles are not new.

**Dictated Stories:** The *dictated story*, for example, is hardly new. It has, for one example, been employed with great success in New Zealand for many years.

A dictated story is exactly what it's name suggests: a story dictated by students, and written down by the teacher. There are many ways to conduct a lesson based on a dictated story, but usually the following steps are involved. First, instruction occurs in a small group with the teacher doing all of the writing. The students' job is to generate the ideas for the story. The teacher's job is to turn that oral language into writing. The lesson begins with the selection of a subject as the focus of the story. This choice can be made by an individual or the group. Once the topic is chosen, it becomes the title of the story and every group member becomes an author.

From this point on in the lesson, decisions must be made

bearing in mind the ability levels of students. With younger students, for example, who are writing about cats, questions such as "Tell me everything you know about cats," may be used to prompt contributions. Once contributions begin, the teacher transcribes them exactly as they are given. The story ends at the completion of a full page, or when the group feels there has been closure on the topic.

Once the story is written it can be used as the focus of the reading lesson. At first, the teacher may read the story back to the students. Rereadings can be done chorally, or by individuals. Skill instruction can readily be built into the reading of the story. Given that the teacher is taking exact dictation, the story may often contain grammatical or other types of errors. During the rereading of the story these issues can be addressed. Stories can be duplicated and passed out to students to take home, and they can be recycled for later use.

**Theme Stories**: This is another example of holistic instruction. In *Learning and Loving It: Theme Studies in the Classroom,*[8] Ruth Gamberg outlined an approach to learning that rejects the traditional approach of relegating knowledge to separate and distinct content areas composed of basic "skills" necessary for the mastery of the material in that area.

Theme studies involve in-depth study of a subject or topic over a period of time. Such study, by definition, involves integrated instruction. A theme study involves much more than reading a story or two about a specific topic. It involves taking an issue and examining it from multiple perspectives. Such a multifaceted approach requires a variety of information sources as well as a variety of instructional activities.

In a theme study the curricular focus is on concepts or ideas versus skills. The study of a specific theme is conducted utilizing traditional skill areas such as reading, writing, and problem solving in math as "tools." Each of these "skill areas" is used to research and explore the topic, relate information about the topic, and complete meaningful activities. For example, Ruth Gamberg describes a theme study for elementary students focusing on housing around the world. Within the study, students gathered information (via reading), and they made models, books, posters, and games to illustrate and share with others what they had learned.

In theme study, topics are chosen not to meet a pre-set

sequence of skills, but rather because they are important to know or because they are of interest to kids. And, because they are important information, students are given a purpose for learning: "The study is not just a thin excuse for teaching children to perfect their word recognition, spelling, punctuation, and so forth. These skills are indeed learned but not through drill. They are learned because they are identified as necessary tools for achieving a purpose.[9]

First and foremost, dictated stories and theme studies place learning in a functional and meaningful context. With theme studies students know exactly what they are studying and why. This focus provides a context in which skills can be developed and utilized. There is no better time to introduce the concept of scale and proportion than when students need to know this for the building of a scale model of a house or classroom. The same holds true for dictated stories. Students are eager to learn the skills necessary for reading their own stories.

Second, theme studies and dictated stories both place a high value on students' interests and motivation. With this as the central tenet of instruction, the teacher can be assured that students will invest the energy and commitment necessary to learn the skills required to complete each activity.

Third, dictated stories and theme studies are active in nature. Within dictated stories, generating a text requires active participation. This activity may not be solely physical in nature, but nonetheless such instruction keeps students actively engaged. The building of a scale model of a structure in a study of housing requires cognitive and physical activity. Physical activity is a critical dimension of learning for all children.

Fourth, theme studies and dictated stories build upon social interaction and cooperative groupwork. Such social interaction serves several instructional purposes. As Ruth Gamberg states:

> Teamwork emphasizes that learning is a social activity and that two or three working well together can produce more satisfying results than one working alone. What one doesn't know, the others may be able to supply. By working together children demonstrate skills for each other. It's a more secure way of learning.[10]

The inclusion of groupwork ensures that students verbally interact during the completion of their projects. Such interaction facilitates learning for everyone involved. Additionally, by working in groups, students develop the social skills necessary in cooperative endeavors. This in and of itself makes theme studies worth the work.

As in reading, the movement to mathematics instruction sensitive to the six principles of learning characteristic of oral language development has gained momentum and many converts in recent years. As with reading, instructional approaches based on these principles are not all new.

**Manipulatives:** The movement to base initial mathematics instruction on manipulatives is an example of an instructional approach sensitive to the importance of having students base their understanding of math concepts on experiences with real objects. In *Mathematics Their Way*,[11] Mary Baratta-Lorton outlines an activity-centered mathematics program for young children that, as its title suggests, is based on direct physical experience with objects and materials in their world.

All activities in *Mathematics Their Way* begin on the experiential level. Manipulatives are used to ''illustrate the concept concretely.'' No symbols are introduced until after students have grasped the basic concept. As Mary Baratta-Lorton states:

> During the beginning stages of concept development, abstract symbolization tends to interfere rather than enhance the understanding of a concept. For this reason, a great deal of this book deals with ideas that develop concepts without the use of any written numerals. Abstract symbolization is only used to label a concept which the child already grasps, never as a ''material'' from which the child is taught a concept.[12]

Lorton's activities employ commercially available materials such as unifix cubes, geoboards, and pattern blocks, as well as ''real'' materials such as shells, bottlecaps and other items that the students themselves are encouraged to collect. These manipulative materials are used as the basic materials for all subsequent instruction.

The rationale behind *Mathematics Their Way* is based on the fact that students' mathematical understanding progresses through three levels or stages. The first, the experiential level, is the focus of Mary Baratta-Lorton's work. For her, this initial level is critical for the development of conceptual understanding of all higher order mathematics. Students' knowledge of mathematical concepts must be firmly grounded in experiences with real world objects. This grounding is important given students' focus on the concrete at this age, but also because such experiences provide the foundation for an understanding of the functional uses of mathematics.

The use of such manipulatives should not be considered to be the sole province of the early elementary grades; quite the contrary. Mary Baratta-Lorton also recommends the use of manipulatives at the *connecting* and *symbolic* levels of mathematical development. As students begin to move to an understanding of symbols and their use and meaning, their understanding of math concepts can be strengthened by the continued use and discussion of the relationship between the concept and the real world. The idea that an abstract concept rests on an understanding of its application in the real world holds for all levels of mathematics, from numbers to beyond algebraic expressions.

**Word Problems:** That the movement toward instruction in mathematics is consistent with the principles of holistic learning can also be seen in instructional approaches currently being employed with word problems. Word problems have been described as a ''perennial bugaboo'' for children, causing so much difficulty that teachers often abandon this dimension of the curriculum.

Students' difficulty with word problems, Marilyn Burns[13] contends, is not their basic computational ability, but rather how they have been taught to solve such problems. The traditional instructional approach presents computing first (adding, subtracting, multiplying, and dividing) via a skills approach, with the hope that students will be able to apply these skills to word problems. The difficulty with this reasoning is that starting with computing removes the functional and meaningful context necessary for the solution of word problems:

> To make sense out of word problems, students need to connect the suitable arithmetic processes to the situations presented in the stories. It's the connecting that causes difficulty. Children can perform computations, but the processes don't make sense to them...[14]

If students cannot conceptualize a word problem in relation to a real world situation, they are not going to be able to decide how to approach the problem.

Successful instructional approaches to teaching students how to deal with word problems, Burns and Richardson state, rest upon having students link the mathematics to real world situations. That is, students must see the function of mathematical computing: "The child needs to see that developing computing skills serves a purpose — that computational skills are the tools for solving problems. Arithmetic needs to be taught in a context that makes sense to the child."[15]

To do this, students must be given opportunities to understand the patterns and regularities that operate in word problems. This situation is no different from the one described in early reading development. Language educators argue that students must be given repeated exposure to the regularities of print, and that it is from these exposures that they will "construct" or generalize the rules and regularities involved.

To facilitate skill in solving word problems, Burns and Richardson say students must be given opportunities to "act out" the situations described in word problems. This acting out can take the form of linking the problem situation to manipulatives, as you would if the problem stated, "Jackie had six cartons of milk ...," or by physically acting out the situation described.

Not only must the situations described in the word problems be made real, students need to have the opportunity to discuss the situation and their thinking about the problem and its solution. Given the opportunity, students actively attempt to identify the patterns and regularities that operate in such problems. Discussion and reflection on the processes and procedures students employ can help facilitate this process.

Mary Baratta-Lorton's approach to the use of manipulatives in early math instruction and Burns and Richardson's approach to teaching word problems are tied directly to the three levels of mathematical development: *experiential, connecting, symbolic.* Instruction, Mary Baratta-Lorton says, must begin with real world objects, with students acting out and using manipulatives to reconstruct the situation. As students grasp the basic concepts, instruction moves on to the *connecting* level. At the connecting level, teachers can act as recorders, connecting the symbols for the concepts with the situation. As Marilyn Burns states: "One way to write the story about Luis and Dianne and the chairs is with words, another way to write the story is $4 + 3 = 7$.''[16] It is only later, after many opportunities for practice, that students will be able to operate independently at the symbolic stage.

Using manipulatives and teaching word problems, in the manner described by Burns and Richardson, are consistent with the principles of holistic learning. First, both instructional practices provide a functional and meaningful link to the real world. They are grounded in activity and interaction with the world and, because of this, students will not lose the focus on the whole in their chase for competence with the parts. Also, both activities draw easily upon social interaction and language use. Working in pairs and groups provides additional opportunities for working with the issues in question. Finally, both instructional approaches are child-centered versus teacher-centered. Both are sensitive to student affect and motivation (they're fun!), and are consistent with the development of mathematical concepts.

## Whole Math Activities: Implementing Change

Our ability to change the current situation in math instruction rests in large measure upon our ability and willingness to make instructional decisions consistent with the principles of learning evident in language development. Instructional activities such as dictated stories, theme studies, manipulatives, and student-created word problems have helped to begin the movement in this direction. The activities that follow are designed to help teachers move from traditional skill-based instruction to instruction more closely aligned with what we know about learning.

How does this ''new'' approach differ from a traditional approach? In a traditional math curriculum, basic geometric shapes, for example, might be introduced to children by having them connect pictures of shapes with the names of those shapes on a worksheet. They might also cut out the shapes and glue them onto paper, making designs and patterns. They will likely have homework sheets that provide additional practice at identifying and naming these same shapes.

Our approach varies signficantly from this traditional one. In a whole math classroom, basic geometric shapes may be introduced and reinforced through an activity such as *Patterns Walk*. (See page 62.)

*Patterns Walk* was designed with reference to the six principles of learning. This activity draws upon the fact that *learning occurs in social settings*. While on a *Patterns Walk,* the teacher or peers guide and encourage children to discover for themselves the variety of geometric patterns that exist in their immediate environment. Discoveries are made and shared in a social setting. The whole class experience is then brought back to the classroom. This initial process of discovery and sharing facilitates learning.

Without a doubt, activities such as *Patterns Walk* beat sitting by yourself at a desk and listing all the triangles found in a picture on page 58! Activities such as *Patterns Walk* are successful because they build upon the fact that *learning is greatly enhanced when the learner is affectively motivated,* that is, when students feel encouraged to take risks and grow in a supportive and comfortable environment. Exploring in and of itself is fun. When students walk and talk together they are free to take risks, make guesses, be "wrong," and reform their ideas. Imagine students in teams of four, excitedly pointing out, sketching, or listing examples of geometry they find as they explore their school or neighborhood. Some suggestions are accepted by the group. Others, for various reasons, don't get recorded. But, the students are working together and are in charge of their own learning; they are making initial discoveries and decisions by and for themselves. This sets a tone and foundation for future lessons in the classroom.

On a *Patterns Walk* students see geometry in use — it has a purpose, whether functional or aesthetic. Roof lines, fences and gates, and the outlines of trees all reflect basic geometry in their lives, and these shapes are related to the function of objects. A *Patterns Walk* around the school or neighborhood provides an opportunity for teachers to link geometry to other aspects of the environment, and creates a *functional and meaningful setting* for learning about math. And the usefulness of math is a major factor in the nature of the learning that takes place.

On the walk, students' observations are dependent upon what they, as individuals, are capable of understanding and relating to at their immediate stage of experience and understanding. Different children see different things, and this activity allows for these differences. If all students began their lesson on page 58, it is possible that many would either not be able to understand much of the page or be well beyond it. Lessons written for the "average" child are limited. We must always be conscious of the fact that *learning is a gradual, developmental process that does not follow a single timeline*. Change, or learning, occurs slowly over time, with different children moving or progressing at different rates.

Our walk and talk is the essence of active learning. We must not forget that *learning is derived from direct experience*. When students touch and see the roundness of a stone or the patterns in a fence they are experiencing the variety, regularity and interconnections between the world and mathematics.

With whole math instruction students' experiences and current knowledge mark the beginning and the foundation of instruction. Exploring and observing the environment by handling objects such as leaves and stones creates a foundation of experience students can draw upon when discussing and writing about patterns and uses of mathematics. Subsequent skills instruction is then enhanced by examples drawn from these holistic "introductory" activities.

Don't limit your class to one *Patterns Walk*. Go again and again. Children's perspectives and knowledge broadens with repeated experiences: *Repeated opportunities for activity and interaction* facilitates learning.

In addition to reflecting the six principles of learning, the *Patterns Walk* is sensitive to, and builds upon, the three stages of numeracy development. Exploring and collecting examples of geometric shapes from the environment are *experiential* by nature. Students on a *Patterns Walk* can manipulate and interact with geometric objects. Connections between these experiences and the language of mathematics can be made in the field or in the classroom when the group discusses and shares their experiences. Students who can identify a geometric pattern, but who do not have a term to describe that shape can, through discussion and interaction with others, acquire the language of mathematics. The movement to the use of symbols is fostered by both the experiences in the environment and the discussion accompanying it. When a teacher draws a triangle on the board after examples of triangles have been collected and examined and discussed, there is a much greater likelihood that students will have the basic understanding necessary to work with this abstraction.

SUMMARY

Whole math instruction should be based upon our understanding of how language ability develops — the basic principles that operate and the stages and changes that occur. Competency in mathematics is facilitated when it is built upon an experiential base tied to functional and meaningful interaction in the real world. The ability to deal with more abstract symbolic forms, like numbers, develops gradually as children construct the patterns governing these forms. Language and social interaction play a key role in this process. It is through discussion and interaction with others, adults and peers, that learning occurs.

# BASIC INSTRUCTIONAL
# METHODS IN WHOLE
# MATH INSTRUCTION

Three basic instructional methods are found in the activities provided in this text: group work, writing, and discussion. These three methods are the key to implementing instruction consistent with the principles of learning and the basic stage of numeracy development.

**Group Work:** A component of all of the instructional activities provided in this text is cooperative groupwork. Social interaction, as we have seen, is a cornerstone of learning. Group work fosters communication and interaction — critical elements of a whole math classroom.

When students are on a *Patterns Walk*, for example, working in groups provides the opportunity for students to share their discoveries, questions, and doubts. This sharing allows not only for student growth in relation to geometric shapes, but also for the development of social skills. Traditional classrooms, with their focus on the individual, do not foster social interaction. Rather, the product of this individual focus is competition, which can lead to an increase in the academic distance between students. The ability of students to communicate effectively with one another and their teachers is the key to success in academic and social development.

Research has shown[17][18][19] that the use of cooperative groups facilitates the development of self-esteem, emotional maturity, social participation, and independent thought, all of which are key elements of a holistic teaching environment. The fact that group work has these additional benefits makes it even more important that it be built into the curriculum.

**Writing in the Holistic Classroom:** The second major component built into the activities provided is writing. It is our belief that students should write about their experiences with mathematical concepts whenever possible. Writing, for our purposes, acts as an umbrella over all the activities in the math curriculum. Language is a tool for learning. Once kids have developed basic language skills, language becomes the major tool for learning new concepts. It is not until they talk about and write about a concept that they understand it.

Depending on the literacy level of the student, some form of written response should be an ongoing aspect of instruction. Putting ideas and concepts into words helps to ensure that such concepts are understood and can be applied in appropriate settings. As the authors of *Using Writing to Learn Mathematics*[20] have noted: "The act of writing gives students the opportunity to formulate, organize, internalize and evaluate concepts."

Facilitating this type of thinking at the primary level can be as simple as requiring students to draw a picture of what they did that day. After a kindergarten class has taken a *Patterns Walk*, they can draw what they have seen. Each child or group can dictate a caption for the page or pages that they have contributed. These illustrations can go into a class *Patterns Walk Book* to be reviewed as a whole class, or enjoyed by individuals during silent reading. This same bookmaking technique works well with older students, too. The descriptions are more elaborate, while the motivation and interest stay just as high.

Other writing activities can be employed by older children. For example, students can maintain a math log or journal in which they keep an ongoing account of their activities. Particularly important for this record is their response to changes in thinking, frustrations, failures, "ah ha's," and the tracking of their development from having no clue to what is going on to a broad conceptual understanding of the concepts at issue. If a *Patterns Walk* is an opening activity for a geometry unit, be sure to go for a walk at the end of your study. Students' writing should provide an excellent assessment of their growth and insight into geometry in their world.

In addition, when students are encouraged to make daily entries in a math log or journal, they are given the opportunity to communicate informally and personally with the teacher. Such entries assist those students who do not wish to risk questions or comments in a large group. Written responses provide an excellent diagnostic and assessment tool both for immediate feedback and for long-range tracking of progress. They build student-teacher relationships. When student comfort levels increase, their written offerings increase.

For all the activities, time should be taken toward the end of each work period for students to reflect orally and in writing on their activity. To facilitate this reflection, students should be encouraged to write in prose form rather than using equations or more traditional mathematical forms when explaining their work. Directions for such reflection might take a form similar to the following:

> Please write about any ideas or strategies used when working today. I especially want to know about any problems you had. You may not have solved them yet, but I would like to know how you have proceeded so far and how you plan to continue. Include the steps you and your partner/group may use to continue and what might change.

As teacher, you must continually model the type of writing you expect from students. Students can easily fall into the pattern of writing simple descriptions of the task or simple statements: "It was fun," or "It was boring." Periodically, the whole class needs to work together to find acceptable ways to put thoughts into words.

**Discussion:** At all grade levels time needs to be taken for discussion and processing of the day's events. At least 20% of class time should be given to group or whole class explanation and processing of the activity. Your role here is to help students frame their thoughts and begin the reflection process. Do not talk at the students and tell them what they have learned.

Putting concepts into words is the cornerstone of learning. Students first experience a new concept through activity in the physical world. With your guidance, via discussion and modeling, students learn to frame their experiences using language. Once they can articulate a concept, either in discussion or writing, they are well on their way to assimilating this new knowledge. Such discussion and processing can act as a pre-writing activity in the math curriculum. Giving students the opportunity to discuss the activities, their actions and feelings, sets the stage for subsequent writing activities. For example, if the topic of a day's lesson is a specific arithmetic operation, such as multi-plication of two or three digit numbers, students can be asked to describe in detail the directions for performing the operation. You can facilitate this discussion with questions such as: "Does this work?" or "What else should be included in the directions?" Once the discussion is completed, students can begin captur-ing such processes in writing, providing further opportunities for learning.

**Student Resource Material:** Children's literature is another source for connecting math with children's interests and imagina-tions. Literature and mathematics share similar purposes. They are about relationships, patterns, classification, problem solving and the beauty that is found in the world around us. They both make our lives richer.

We have included some titles of children's books or other resources that we think will further enhance and broaden your stu-dents' experiences when exploring the activities.

**Limitations and Provisos:** The basic principles discussed here must be understood in the light of the real classroom. Certain provisos hold for the activities provided in the next section. Vari-ations and adjustment must occur in instructional methods across age and developmental levels. Reading and writing tasks must necessarily vary for children of different ages. First of all, lan-guage development limits the options that are open to you. Youn-ger children, who are still working with gross motor control and basic literacy skills, have a hard time with extensive writing assignments. Further, the nature of oral language activities varies across ages and classrooms. Less in-depth discussion will likely occur with younger students. Activities in primary classrooms involve more exploring and less talking — but older children need to explore, too! Older students can take responsibility for a greater amount of the processing of tasks, and language can and should be brought to bear with these students.

Group work also operates within certain parameters. Cooper-ative groups of four are very different with kindergartners. Fur-ther, more teacher management is required with young children, with these teachers taking a more active role in group work, writ-ing tasks and discussions and offering their younger students a greater degree of mediation.

# WHOLE MATH ACTIVITIES

# FOR PRIMARY AND

# ELEMENTARY CLASSROOMS

The activities given in this text are designed to provide a starting point for teachers who wish to utilize language and holistic instruction as a tool for the development of mathematical concepts.

The activities are organized and presented by major strand and recommended age level(s). The strands represent the general categories of mathematics: measurement, number, geometry, logic, and probability and statistics. These general categories represent the units and topics most commonly found in math textbooks.

Each activity is presented under six major headings: *suggested ages; content; materials; why?; how?;* and *thinking it over*. These headings establish if the activity is suitable for your students, highlight the nature and objective(s) of the activity, and suggest how the activity may be introduced and developed.

As you examine and experiment with the suggested activities, you will undoubtedly notice that certain ones could be designated as relating to another strand such as logic instead of measurement. Do not be alarmed. You're right; certain activities involve several areas of focus and prerequisite skills. What we have done is place the activity in the area *we* felt was most appropriate.

As you examine the activities and consider how they might be employed in your classroom, don't lose sight of the principles and rationale implicit in them. It is our hope that these activities will spur your thinking about math instruction in particular

and holistic instruction in general and that the excitement you
see in the eyes of your children will be your reward.

BIBLIOGRAPHY

1. Steen, L. A. "Pattern." *On the Shoulders of Giants: New Approaches to Numeracy*. Edited by L. A. Steen. National Academy Press, 1990, p. 15–31.
2. Pimm, D. *Speaking Mathematically: Communications in Mathematics Classrooms*. Routledge Chapman and Hall, 1989.
3. { Halliday, Michael A. *Explorations in the Functions of Language*. Edward Arnold, 1977.
    Halliday, Michael A. *Learning How to Mean: Explorations in the Development of Language*. Edward Arnold, 1975.
4. Piaget, J. *The Origins of Intelligence in Children*. International Universities Press, 1966.
5. Inhelder, B. and J. Piaget. *Early Growth of Logic in the Child*. Norton, 1969.
6. Piaget, J. *Success and Understanding*. Harvard University Press, 1978.
7. Burns, M. and K. Richardson, "Making Sense Out of Word Problems" *Learning Magazine*. January 1981, p. 26-32.
8. Gamberg, R., W. Kwak, M. P. Hutchings, and J. Altheim. *Learning and Loving It: Theme Studies in the Classroom*. Heinemann Educational Books, Inc., 1988.
9. ibid. p. 12.
10. ibid. p. 38.
11. Baratta-Lorton, M. *Mathematics Their Way*. Addison-Wesley, 1976.
12. ibid. p. 14.
13. Burns, M. and K. Richardson, "Making Sense Out of Word Problems." *Learning Magazine*. January 1981, p. 26–32.
14. ibid. p. 27.
15. ibid. p. 28.
16. ibid. p. 29.
17. Johnson, D. and R. Johnson. "Cooperative, Competitive and Individualistic Learning." *The Journal of Research and Development in Education*. 1978, *Vol. 12*, (1).
18. Johnson, D. and R. Johnson and D. Anderson. "Relationship Between Student Cooperative, Competitive and

Individualistic Attitudes and Attitudes Toward Schooling." *Journal of Psychology*, 1978, 100, p. 183–199.

19. Johnson, D. and R. Johnson. *Learning Together and Alone: Cooperative, Competitive and Individualistic Learning*. Prentice-Hall, 1987.

20. Nahrang, C. L. and B. Peterson. "Using Writing to Learn Mathematics." *Mathematics Teacher*, September 1986, p. 461–465.

# MATRIX SHOWING MAIN FOCUS BY STRAND

| ACTIVITY | LOGIC | GEO-METRY | MEAS-URE. | NUM-BER | PROB./STAT. |
|---|:---:|:---:|:---:|:---:|:---:|
| 1. Comparing & Ordering | X | X | X | X | X |
| 2. Similarities & Differences 1 | X | X | | | X |
| 3. Similarities & Differences 2 | X | X | | | X |
| 4. Venn Diagrams | X | | | | X |
| 5. Co-op Logic: What's My Line? | X | | | X | X |
| 6. Co-op Logic: What's My Number? | X | | | X | X |
| 7. All Sorts of Stuff | X | | | X | |
| 8. Patterns Walk | X | X | | X | |
| 9. Investigating Blocks | X | X | | | |
| 10. Over the Wall | X | X | | | |
| 11. Pattern Block Creations | X | X | | | |
| 12. Grab Bag | | X | | | |
| 13. Measurement Tools | X | | X | X | |
| 14. Body Measurements | | | X | X | |
| 15. Body Scale | | | X | X | X |
| 16. Classroom Model | X | X | X | X | |
| 17. Time Lines | | | X | X | |
| 18. Calendar Daze | X | | X | X | |
| 19. Measurement Investigations | X | | X | X | |
| 20. Numbers All Around Us | | X | | X | |
| 21. Estimation | X | | | X | |
| 22. Number Posters | | | | X | |
| 23. Number Plays | | | | X | |
| 24. Headlines | | | | X | |
| 25. Hundred Chart | | | | X | |
| 26. The Land of Doy | X | X | | X | |
| 27. Student Statistics | X | | | X | X |
| 28. Burning Questions | X | | | X | X |
| 29. Mystery Graphs: Open Bar | X | | | X | X |
| 30. Mystery Graphs: Natural Phenomena | X | | | X | X |
| 31. Mystery Graphs: The Plot Thickens | X | | X | X | X |

* Shaded cells represent the major focus of the activity

. . . . . . . . . . . . . .

# 1. COMPARING AND ORDERING

*Ages 5–8*

*Content*
logic, geometry, measurement, number, probability and statistics

*Materials*
common objects, yarn, adding machine tape

*Why?*
- Establish the basic concepts of comparing and ordering and the vocabulary associated with these concepts.

*How?*
Begin by having partners determine who is taller by comparing their respective heights. Students can make this comparison simply by standing next to each other, or with yarn or adding machine tape cut to their heights.

*Step 1:* In groups of four have students establish an order of their heights from shortest to tallest. Upon completion of this ordering, have students draw a picture of the group showing themselves in this order.

Take time at the end of this activity to introduce the terms that are most commonly used to describe comparisons of height or length: *small, smaller, smallest; tall, taller, tallest.* Have students frame responses to your questions utilizing these terms.

*Step 2:* Have all students form a circle and move within the circle to create an ordering of all the students from shortest to tallest. When the activity is completed, the shortest will be beside the tallest. This ordering is accomplished by pairs of children in the circle facing each other. They must decide who is taller, who is shorter, and then move one place in the appropriate direction, say, left for taller, right for shorter. Eventually they create an ordering of themselves from shortest to tallest. At the completion of this activity, discuss the terminology commonly used to compare and order objects by height or length. Have students brainstorm a list of these terms, and post the list in the room.

### Progression 1

When students are comfortable with the concept of comparison and order, shift the focus from them to objects in the room. Compare and order various pairs or groups of objects (for example a book, a pencil; a piece of chalk, an eraser) based on height or length. Next, introduce new objects into the set, for example by adding additional objects to the first pair: a ball, an orange, and a piece of string.

Have students work with partners to collect, order, discuss, and draw or write about their own sets. Explore the many possible variations. To guide their explorations, ask questions such as: ''Are there other ways these objects can be compared or ordered? Can we compare their weight (mass)? color? age? cost? What other types of comparisons are possible?'' Once this discussion is complete, have students create pictures and graphs to represent the various orderings and comparisons. Post them around the room. You can use these pictures to further the discussion. Encourage students to write captions or explanations to go with each picture.

### Progression 2

Later, when basic measurement concepts and procedures have been introduced (rulers or metresticks), students can take measurements and compare the numerical value of the results. This is often a difficult task for beginners, so lots of non-numerical exploration should come first. They must be quite comfortable and confident with the use of manipulatives before they move to numerical abstractions.

*Thinking It Over*

It is important to use the children's bodies and objects they choose as initial manipulatives for these exercises. In addition, constant repetition is important for them to become comfortable with the process and the concepts associated with comparing and ordering objects by height or length. Repetition also helps students to develop the language used to compare and describe objects. Model ways for them to interpret and explain what is happening, and then give lots of time for them to do the same.

STUDENT RESOURCE MATERIAL

Hoban, Tana. *Is It Red? Is It Yellow? Is It Blue?* Greenwillow, 1987.
_____. *Is It Larger? Is It Smaller?* Greenwillow, 1985.
*My First Look at Sizes.* Kindersley, 1990.
Parnall, Peter. *Feet!* Macmillan, 1988.
Pienkowski, Jan. *Sizes.* Julian Messner, 1983.
Spier, Peter. *People.* Doubleday, 1988.

. . . . . . . . . . . . . . .

## 2. SIMILARITIES AND

## DIFFERENCES: I

*Ages 5–7*

*Content*
geometry, logic

*Materials*
pencil, paper, chalkboard, manipulative alphabet letters

*Why?*
- Develop abilities to sort and classify by simple attributes.
- Develop ability to think logically.

*How?*

*Step 1:* Begin the activity by choosing two letters, perhaps from the beginning of two children's names. Ask: ''Can you look at these two letters and tell me what is the same about them?'' For example, responses to the letters B and H could be: ''They're both capital letters'' (simple, but acceptable), or ''They both have straight lines.''

As students contribute responses, list them on the board under the heading entitled *Same.* If necessary, offer hints and suggestions to guide students to the geometric properties of the letters.

Take the same approach with *Differences,* such as ''One doesn't have any curves.'' Again take students' comments and list them on the board for future reference.

*Step 2:* After this process has been modeled, children work in pairs choosing the first letters from their names and noting similarities and differences, recording them with pencil and paper, if possible. When this process is completed, the class reconvenes for a general discussion of the activity and their findings.

*Progression*

*Step 1:* Once students have an understanding of the basic process, ask students to sort all the letters of the alphabet into two categories. Have them explain how they chose their categories (cut-out or plastic letters can be used as manipulatives).

*Step 2:* Next, sort the letters into three or four categories. Again have them explain their choices, and share their lists of similarities and differences with the class. Have students draw or write about their categories. Student work can be posted or placed in a classroom booklet of group projects.

*Step 3:* Have partners or groups create Venn Diagrams (page 00) on construction paper with the attributes of the letters as the framing statement. For example, one of the guiding statements could be *letters with straight lines*, another might be *letters with curved lines*. As with all activities, encourage diversity and variety.

*Thinking It Over*

Any number of everyday objects from students' lives can be used to generate the logic and skills inherent in this activity. Using the alphabet, specifically letters related to them even in the simplest way, helps to integrate basic math and thinking skills with personal interest. Again, when the curriculum has a direct connection with the children's world, they are more likely to integrate new learning into their present bank of knowledge and skills.

STUDENT RESOURCE MATERIAL

Hoban, Tana. *Is It Red? Is It Yellow? Is It Blue?* Greenwillow, 1987.
*My First Look at Sorting.* Dorling Kindersley, 1991.
Reid, Margarette S. *The Button Box.* Dutton, 1989.
Sis, Peter. *Beach Ball.* Greenwillow, 1990.

. . . . . . . . . . . . . . .

# 3. SIMILARITIES AND

# DIFFERENCES: II

*Ages 8–10*

*Content*
geometry, logic, probability and statistics

*Materials*
Any geometric manipulatives (such as pattern blocks, attribute blocks, geoblocks) or commonly found objects such as keys, bottle caps, etc.

*Why?*
- Develop abilities to sort and classify geometric shapes by their basic attributes and properties.
- Reinforce basic geometric vocabulary.
- Strengthen abilities to identify patterns and think logically.

*How?*
Students work in pairs or groups with two distinct blocks (for example a square and a trapezoid). Have students generate a list of the similarities and differences between these two shapes. When they have exhausted the possibilities, stimulate a class discussion of the findings.

*Progression 1*
Once this process and concept has been established, give students a collection of geometric objects with the instruction: "Find a way to place these objects in just two categories." You can discuss these categories with students, or they can write about them in their journals.

*Progression 2*

Challenge students to find different ways to categorize the same objects. As before, a discussion and a written response of the findings should follow.

*Progression 3*

Move to three, four or more categories depending on students' abilities. Explore various ways to record or graph the results (see *Venn Diagrams* and *Mystery Graphs*).

*Thinking It Over*

Again, a critical part of this seemingly simple activity, at whatever level, is the discussion that ties together students' processes and observations. Encourage your students to express their thoughts out loud as they make decisions and explain why they have chosen a particular move, so that language becomes the primary tool for organizing thoughts and getting work done. Also, emphasize and encourage variety and creativity.

STUDENT RESOURCE MATERIAL

Parnall, Peter. *Feet!* Macmillan, 1988.
Reid, Margarette S. *The Button Box*. Dutton, 1989.
Sis, Peter. *Beach Ball*. Greenwillow, 1990.
Spier, Peter. *People*. Doubleday, 1988.

. . . . . . . . . . . . . .

## 4. VENN DIAGRAMS

*All ages*

*Content*
logic, number, probability and statistics

*Materials*
paper, pencils or pens, yarn

*Why?*
- Develop the ability to sort and classify based on similarities and differences with a strong emphasis on language.
- Develop understanding of relationships among ideas, concepts, people, and objects

*How?*

*Step 1:* Introduce the process of sorting and categorization by starting with sorting the children themselves. Have students form a large circle. Choose students to move into the center based on one unspoken characteristic: those with belts, or earrings, or curly hair, or buttons. All students then guess by what common characteristics those in the center were chosen.

*Step 2:* After modeling the process a few times, have the student or students who guessed correctly select the next members to enter the circle. Or have students take turns: they can secretly select a characteristic and choose which students move into the center.

*Progression 1*

After students have had many experiences with categorization by one characteristic, move to using yarn circles on the floor of the classroom instead of the human circle. (This is a small progression toward symbolic abstraction.)

After more experience with sorting single characteristics into one circle of yarn, introduce a second overlapping circle to create intersecting sets. For example:

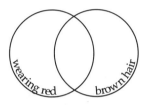

Rather than have students guess characteristics at this point, label the circles and ask students to place themselves within the yarn circles, depending upon their personal attributes.

In this example focusing on brown hair and red clothing, the class would end up with some students in the brown-hair circle only, some in the red-clothing circle only, and several standing in the intersection of the sets — that is, in the overlap between the circles. In addition, some would be in neither of the circles. They would stand outside.

Explore other implications of classification suggested by Venn diagrams. Sometimes circles don't intersect, as with brothers and sisters. Or, on the other hand, sets can be completely included within others. Exploring logical connections can go into all areas of interest.

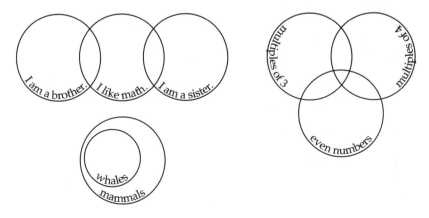

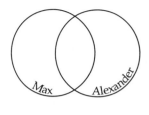

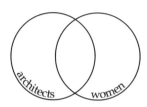

*Progression 2*

After students have physically explored the Venn/set concept, post large, handwritten Venn diagrams for them to respond to. Students write their names in the one appropriate place on each diagram. Just as their bodies can only be in one place at one time, their names can go in only one place on each diagram. As before, you will need to guide students in their responses to the diagrams and in interpreting the results.

*Progression 3*

Have students develop their own diagrams and frame statements. Children can post their diagrams and read and explain them to the class. Leave them up for a time so all students can fill in their names or initials. You may also want to encourage your students to post Venns at home in a busy spot for the family and guests to fill in. Everyone gets involved.

*Progression 4*

Once the diagrams have been filled in — the data has been collected — it's time to interpret your findings. This is where Venn diagrams present their greatest potential.

  As a class, choose one diagram to work with. Begin with a broad question: ''Using the information on Sarah's diagram, can you make one interpretation or observation about our class?'' Responses might be: More people have dogs than cats. Seven people have dogs *and* cats. Twelve people don't have dogs or cats. These are simple, factual interpretations of the data. This is an important skill for students to develop. It's also a safe way to begin. After the class has generated observations about what is known, ask: ''What do we not know from looking at this data?'' or, ''What may be misleading about this information?'' For example, you might say that this diagram shows that ''More people like dogs than any other animal.'' Challenge them to explain why this statement cannot be proven by the information contained in the diagram. You might even challenge them to make misleading statements based on the data. Or, for more experienced groups, challenge them to make opposing statements based on the same information. These kinds of questions ask students to go beyond what is immediately available in the data. This is a beginning step to a crucial survival skill of analyzing and interpreting information.

*Thinking It Over*

Venn diagrams challenge the learner in several ways. On an immediate level, students work to master the logic and language skills required to deal with set and class inclusion tasks, which help develop their understanding of the logical relationships between ideas and objects in their lives. We are all dependent upon language to define and describe logical activity. These simple circles help children develop this critical connection between language and logic. As they are challenged to create their own Venns, they learn to work with language and its interdependence on logic. Yet the greatest potential of Venn diagrams comes when students analyze and interpret data contained in or suggested by the diagrams.

Interpretation of the diagrams can lead to interesting knowledge and awareness for everyone about one another and the world beyond the classroom.

STUDENT RESOURCE MATERIAL

Froman, Robert. *Venn Diagram*. Thomas Y. Crowell Publishers, 1972.
MacCarthy, Patricia. *Ocean Parade: A Counting Book*. Dial, 1990.
Giganti, Paul. *How Many Snails?* Greenwillow, 1988.

## 5./6. COOPERATIVE LOGIC:

## WHAT'S MY LINE?

## WHAT'S MY NUMBER?

*All ages*

*Content*
logic, number

*Materials*
Sets of clue cards, hundred charts (see end of this activity), various objects such as beans and blocks to use as markers and manipulatives. You may want to print up each set of clues (card stock is best) and put individual sets into labeled envelopes. This simplifies the distribution and management of the activities.

*Why?*
- Learn cooperative group skills.
- Develop problem-solving ability.
- Develop facility with vocabulary and syntax.
- Review and reinforce various mathematics skills.

*How?*
The object of this activity is for students to solve the problems posed on the various clue card sets. In the *What's My Line?* section students determine the appropriate order of the individuals named on the cards based on the clues provided. In *What's My Number?* students determine which number (or numbers), between 1-10, 1-50, or 1-100, satisfies all of the clues provided on the clue cards.

*Step 1:* Have students work in groups of 2, 3, or 4. If they are readers, each student should receive one or more clue cards from a set. If students are not at an independent reading level, you can read the clues and the groups decide on placement of the individual in the line or number on the chart based on your statements.

**Important:** Students take responsibility for their own clue(s). They read their card to the rest of the group, or tell in their own words what they think their clue(s) means. They do not, however, allow others in the group to look at or hold their card(s).

*Step 2:* The group discusses the clues and decides how to arrange the individuals or chooses which number or numbers are best so that all clues are satisfied. When all members of the group are satisfied with the solution, the clue cards are put down face up for every member to see, and the group then decides if the choices they made were correct. If they are not satisfied, they need to determine how and where they got off track.

*Step 3:* After all groups have worked with a particular card set, bring the entire class together to discuss strategies. It is important to recognize that each group, indeed each person, may have a different way of perceiving and solving the problem. (Note that a few of the problems have more than one correct answer.)

At the beginning, place your emphasis on the interaction of group members. The first questions should be similar to: "How did your group manage together? How did you get along? How did you decide to accomplish your task? Were there any problems? How did you solve them?" This line of questioning shows students that their interaction — the cooperative group process — is the first priority in the lesson. The answers they develop are secondary.

When discussing an activity, whether around interaction of the group or for answers, find ways to emphasize the positives, particularly when it is determined that groups might have different answers: "Why can't there be two different answers? If you think this answer should be changed, can you convince others? Do you remember what happened during the activity that led you to this answer? Let's review our clues."

Part of the expected outcome of this task is for students to work together to discover a common solution. Be sure to stress that positive interaction among group members is as important to group success as finding a workable solution to the problems at hand.

*Step 4:* Independence. After the class has worked through the clue sets provided here, have individuals or groups make up card challenges for others. Keep a file box for students to add to or take from in their free time.

*Thinking It Over*

A major purpose of *Cooperative Logic* is to give students opportunities to solve problems within groups. It is important for them to work out their social as well as academic challenges as often as possible without direct intervention from you. This will be difficult for many at first, but students quickly come to enjoy the responsibility and comradeship. They are developing invaluable skills for their future.

To ensure that the groups work successfully, please keep in mind the following issues:

**1. Ownership:** Make sure that students stick to the rule that each group member has ownership over her or his particular clue card(s). This helps to ensure each child has a say in the group, is not overpowered, and does not drop out. Card ownership also helps to ensure that no child dominates in the group.

**2. Expect a noisy room:** When this activity is in full swing there is lots of discussion. It can be noisy but, if the activity is working, it's good noise.

One important goal of this activity is for students to be in charge of their own learning. As they become more independent, take the opportunity to move around the room and observe your students from a different perspective. You are no longer directing the lesson from the front of the class; but rather you are a facilitator, or aide, for groups that might get stuck or need a little nudge.

**3. Cooperative learning groups take time:** Students and teachers both need time and experience to get used to this new way of learning and management. Students may be reluctant at first to express themselves. Some will need to develop the skills and confidence to contribute to the group. Others may want to dominate the discussions and will need to learn to share. Many students used to a traditional, competitive classroom environment need time in the cooperative experience to develop a trust that their peers and their teacher are working together to solve problems. For students new to cooperative learning, having a specific product as an outcome helps to focus the group task. Having a clearly defined task allows students to tell when they have completed that task, and provides a greater opportunity for success.

**4. Right answers:** As students become more accomplished and confident in cooperative groups they can be expected to work on far more open-ended problems; not just in mathematics, but in all areas of the curriculum. They might work together to design a school, write a play, or present a report as a team. The cooperative problems presented here, however, have definite solutions. This has been done to limit the number of things that might cause groups to falter.

STUDENT RESOURCE MATERIAL

Fraser, Sherry. *Spaces.* Berkeley, CA: EQUALS Project, Lawrence Hall of Science, University of California, 1982.

Erickson, Tim. *Get It Together: Math Problems for Groups, Grades 4-12.* Berkeley, CA: EQUALS Project, Lawrence Hall of Science, University of California, 1989.

Goodman, Jan. *Group Solutions.* Berkeley, CA: GEMS Project, Lawrence Hall of Science, University of California, 1992.

**Note:** *For student use, enlarge each activity sheet by 130-150%.*

WHAT'S MY LINE?: DIRECTIONS

- Lines A-D are the easiest to solve. They have four characters and four clues each. **Note:** Use the blank template for B, C, D.
- Lines E-H get a bit more difficult. They have five characters and six clues each. **Note:** Use the blank template for E, F, G, H.
- Cut out individual clues and name tokens. Each group gets a set of clues and tokens for each problem.
- There is no order to the clues in each set. They are lettered and numbered so you can make sure each set is complete with no duplicates.
- Clues are passed out to every member of the group. Depending upon the size of groups, students may have one or more clues each.

- Each students holds a clue(s). The clue is read or paraphrased to other group members but they cannot look at it.
- The name tokens are placed on the table or floor so that all members can see and manipulate them.
- Students then read or explain their clues to one another working together to decide upon a correct sequence for the name tokens.
- When all members of the group are satisfied that their cooperative decision is correct, they put their clues down so that everyone can read them. They check for agreement once again.
- If, upon further examination, a group decides that their original line order is incorrect, they need to retrace their steps and discuss where they got off the track and what they should have done. (This can be critical to the learning process.)
- Be careful of literal interpretations of clues. For example: "Alice is in front of Beto" could mean she is directly next to him or a few places ahead of him.
- When a group decides it is finished with a problem, they may go on to another one or wait for the whole class to process the same problem together.
- Several of the problems have more than one logical solution. The intention is to promote discussion. Have students give logical arguments for their decisions and help them understand that there can sometimes be more than one correct solution.
- No answers are given. You and your class work together to decide if you think your solution(s) works. The final progression is for students to use the blank template to create their own "What's My Line?" problems.

WHAT'S MY LINE?: CLUES

Note: If you wish, substitute names of children in your class.

**B:**
1. Daryl is before Beto.
2. Alice is after Beto.
3. Beto is before Carlos.
4. Carlos is not last.

**C:**
1. Daryl is behind Alice.
2. Alice is in front of Carlos.
3. Beto is between Daryl and Carlos.
4. Daryl is not last.

**D:**
1. Carlos is before Daryl.
2. Alice is before Daryl.
3. Carlos is behind Alice.
4. Beto is between Carlos and Alice.

**E:**
1. Daryl is in the middle.
2. Beto is not next to Erin.
3. Alice is next to Erin.
4. Beto is only next to Carlos.
5. Beto is ahead of Alice.
6. Erin is right behind Daryl.

**F:**
1. Carlos is after Daryl.
2. Beto is after Carlos.
3. Erin is before Beto.
4. Alice is between Daryl and Carlos.
5. Alice is second in line.
6. Erin is not first.

**G:**
1. Erin is not last.
2. Carlos is not first.
3. Beto is in the middle.
4. Alice is behind Beto.
5. Daryl is before Beto.
6. Alice is next to Carlos.

**H:**
1. Erin is behind Beto.
2. Carlos is between Alice and Beto.
3. Daryl is not last.
4. Carlos is before Beto.
5. Daryl is not first.
6. Erin follows Daryl.

WHAT'S MY NUMBER?: DIRECTIONS

**Note:** The directions here are similar to "What's My Line?" The procedure for distributing the clues in this activity are the same, and so is the use of the blank template.

- Each group has one copy of the Hundred Chart. They may also use markers (beans, tabs) to select or eliminate numbers as they work through their clues.
- Numbers A-D have four clues each. Students work with the numbers 1-10 on the chart.
- Numbers E-H are between 1-50 and also have four clues each.
- Numbers I-L are between 1-100 and have six clues each. Each of these problems can be solved using the first four clues only. The last two clues marked *extra* can be used with the others

as just two more clues, or set aside and used only in case the group gets really stuck.

- Just as with "What's My Line?," students read only their own clues but share the Hundred Chart and work together to find the number.
- When they are confident that they have found a number that satisfies all their individually held clues, they put the clues down so that everyone can read them. The group then makes a final assessment of their solution.
- They then move to the next set of clues, wait for whole class discussion, or create their own set of clues for others to solve.
- Remember, there may be more than one correct solution.

## WHAT'S MY NUMBER?: CLUES

**Note:** If you wish, substitute names of children in your class.

### B: Bert
1. It is less than 5 + 5.
2. It is closer to ten than zero.
3. It has fewer than 5 letters.
4. It is an odd number.

### C: Carmen
1. It is an even number.
2. It is not 4 + 4.
3. It has only one digit.
4. It is greater than 1 + 3.

### D: Duane
1. It has fewer than 5 letters.
2. It is not 2 + 3.
3. It is less than 9.
4. It is greater than 3.

### E. Elvis
1. Its digits add to an even number.
2. It does not have a 4 as one of its digits.
3. It is greater than 20.
4. The difference between its digits is 2.

### F. Flora
1. Her number is less than 40.
2. You get to her number by counting by fives.
3. Flora's number is greater than 20.
4. One of the digits is a 2.

### G. Glenda
1. The difference of the digits is greater than 4.
2. Her number is not even.
3. Nine is not one of the digits.
4. Both of the digits are odd.

### H. Hank
1. Hank's number is a multiple of 4.
2. The difference between its digits is less than 4.
3. It is a multiple of 3.
4. The difference between its digits is an even number.

## I. Ilsa

1. The sum of the digits is 7.
2. Ilsa's number is even.
3. The digit in the tens place is less than the digit in the ones place.
4. The difference between the digits is greater than 1.

*Extra*

5. Ilsa's number is a square.
6. One of the digits is a 1.

## J. Jesse

1. Jesse's number is a multiple of 5.
2. The sum of the digits is greater than 8.
3. Six is not one of its digits.
4. The sum of its digits is a prime number.

*Extra*

5. The digit in the tens place is an even number.
6. It is not a multiple of 10.

## K. Kate

1. It is not evenly divisible by 6.
2. The sum of its digits is even.
3. It is a multiple of 4.
4. The sum of its digits is a multiple of 7.

*Extra*

5. The difference of its digits is 2.
6. It is greater than 50.

## L. Lee

1. The numeral 1 is not one of its digits.
2. One of its digits is a multiple of the other digit.
3. It is a multiple of 3.
4. It is greater than 7 squared.

*Extra*

5. The sum of the digits is 9.
6. It is an odd number.

# WHAT'S MY LINE? A
## 4 CHARACTERS
## 4 CLUES

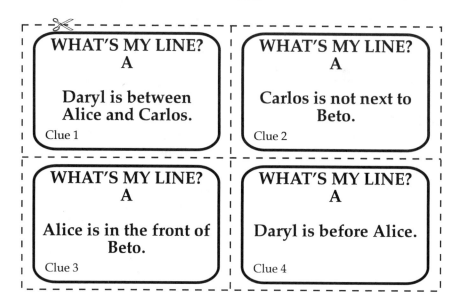

**WHAT'S MY LINE? A**

Daryl is between Alice and Carlos.

Clue 1

**WHAT'S MY LINE? A**

Carlos is not next to Beto.

Clue 2

**WHAT'S MY LINE? A**

Alice is in the front of Beto.

Clue 3

**WHAT'S MY LINE? A**

Daryl is before Alice.

Clue 4

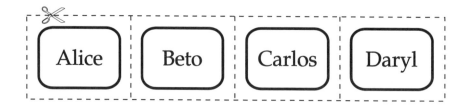

Alice | Beto | Carlos | Daryl

# WHAT'S MY LINE?

**WHAT'S MY LINE?**
_____

Clue 1

**WHAT'S MY LINE?**
_____

Clue 2

**WHAT'S MY LINE?**
_____

Clue 3

**WHAT'S MY LINE?**
_____

Clue 4

**WHAT'S MY LINE?**
_____

Clue 5

**WHAT'S MY LINE?**
_____

Clue 6

See if some friends can figure out your line.

# WHAT'S MY NUMBER? A
## ALMA

**WHAT'S ALMA'S NUMBER?**

It is less than 8.

Clue 1

**WHAT'S ALMA'S NUMBER?**

It is an even number.

Clue 2

**WHAT'S ALMA'S NUMBER?**

Her number is greater than three.

Clue 3

**WHAT'S ALMA'S NUMBER?**

It is not six.

Clue 4

## Make up your own
# WHAT'S MY NUMBER?

### HUNDRED CHART 1-___

WHAT'S _____
NUMBER? _____

Clue 1

WHAT'S _____
NUMBER? _____

Clue 2

WHAT'S _____
NUMBER? _____

Clue 3

WHAT'S _____
NUMBER? _____

Clue 4

WHAT'S _____
NUMBER? _____

Clue 5

WHAT'S _____
NUMBER? _____

Clue 6

See if some friends can figure out your number.

# HUNDRED CHART

| 1 | 2 | 3 | 4 | 5 | 6 | 7 | 8 | 9 | 10 |
|---|---|---|---|---|---|---|---|---|---|
| 11 | 12 | 13 | 14 | 15 | 16 | 17 | 18 | 19 | 20 |
| 21 | 22 | 23 | 24 | 25 | 26 | 27 | 28 | 29 | 30 |
| 31 | 32 | 33 | 34 | 35 | 36 | 37 | 38 | 39 | 40 |
| 41 | 42 | 43 | 44 | 45 | 46 | 47 | 48 | 49 | 50 |
| 51 | 52 | 53 | 54 | 55 | 56 | 57 | 58 | 59 | 60 |
| 61 | 62 | 63 | 64 | 65 | 66 | 67 | 68 | 69 | 70 |
| 71 | 72 | 73 | 74 | 75 | 76 | 77 | 78 | 79 | 80 |
| 81 | 82 | 83 | 84 | 85 | 86 | 87 | 88 | 89 | 90 |
| 91 | 92 | 93 | 94 | 95 | 96 | 97 | 98 | 99 | 100 |

# 7. ALL SORTS OF STUFF

*All ages*

*Content*
logic, number

*Materials*
Word Problems on clue cards (see end of this activity). A variety of manipulatives to serve as representational objects: colored blocks or pieces of paper, even the students themselves. (If you can take the time, have students use construction paper or other materials to create the fruits and birds to be used in the sorting. These representational items are less abstract than blocks). A piece of paper with a drawing of a bowl or wire can serve as the working area.

*Why?*
• Develop reasoning ability using basic mathematics language.
• Provide practice with creating and solving word problems.

*How?*
This activity contains two categories of things to sort: fruit and birds. Each category contains three successive levels of difficulty. The first level is relatively easy and is intended for younger students, or as an introduction for older students. Each level contains ten sets of statements, which we call *sorts*. Either you or your students write the statements for each sort separately on cards. Individual sets of cards can be placed in labeled envelopes. The object of the activity is to frame a question or questions based on the statements in each sort.

For example, statement 1 for bowl A, states: There are 3 apples. There are 7 pieces of fruit. Based on the major opening statement, we also know that each bowl contains apples and oranges. A question such as *"How many apples are there in bowl A?"* focuses student attention on "sorting out" the number of apples in this bowl.

Students answer the question generated. In this sort, for example, another question that can be asked is: "How many oranges are there?" The answer, of course, is that there are 4 oranges in the bowl. Other questions arising from the statements might include: Are there more apples or oranges? How many more oranges are there than apples? Although the questions seem simple, the logical reasoning required to frame them provides students with practice at creating and solving word problems.

*Step 1:* Introduce the activity to the whole class and model the procedure, that is, frame key questions and find the solutions to them. Be sure to use manipulatives as you model the process required from students. When students are framing questions and answers, encourage them to use the manipulative that represent the objects — fruit, birds — in the statements, especially as they progress to more difficult sorts.

*Step 2:* Have students work in groups to read and discuss each sort. (Non-readers may have statements read for them). The groups' task is to frame at least one question and answer for each sort. Asking them to create two or more questions for each sort helps to stretch their logic and language skills.

As we have pointed out, more than one question can be generated given the clue statement in each sort: For example, bowl B, statement 5 says: There are 9 pieces of fruit. There are 3 more apples than oranges. Students might generate questions such as: "How many oranges are there? How many apples are there? What is the relationship between the number of each kind of fruit? What are the fractional equivalents of each kind of fruit?"

Obviously, the depth of language and logic explored depends upon the development and skill of students. You will find that as you work toward the third level of sorts that the statements generate multiple solutions. Don't miss this opportunity to encourage arguments and discussion among your students. Continue to ask: "Are there any more solutions? How do you know? How can you show me?"

Because answers depend on the questions generated, no answers are given here.

## Progression

Students develop their own sorts and share them with other groups. They will choose very creative subjects to sort. It's fun.

## Thinking It Over

This activity provides students with an opportunity to frame questions based on given information, and then to find answers to those questions. These skills are basic to the solution of word problems and even more basic to solving problems in everyday life.

The importance of using manipulatives in the solution of the problems cannot be overemphasized. Too often we ask students to deal with abstract and decontextualized problems without access to concrete objects. Allowing students to make sense of the problem using the manipulatives is the first step toward competency with word problems.

STUDENT RESOURCE MATERIAL

Hoban, Tana. *Push, Pull, Empty, Full: A Book of Opposites.* Macmillan, 1972.
Pluckrose, Henry. *Sorting.* Franklin Watts, 1988.

BOWLS OF FRUIT: CLUE STATEMENTS

A. Major Statement: *There are oranges and apples in each bowl.*
**Note:** All students need to be familiar with this statement, which is the base for all the following clues.

BOWL 1:   There are 3 apples. There are 7 pieces of fruit.
BOWL 2:   The bowl has 2 oranges. It has a total of 8 pieces of fruit.
BOWL 3:   There are 4 oranges. The total number of fruit is 10.
BOWL 4:   There are 8 apples. There are 10 pieces of fruit.
BOWL 5:   There is the same number of apples as oranges. There are 10 pieces of fruit.
BOWL 6:   The number of oranges and apples add up to 9. There are 8 apples.
BOWL 7:   There are 5 more oranges than apples. There are 7 oranges.
BOWL 8:   There are 5 apples. There are 3 more oranges than apples.
BOWL 9:   There are 4 more apples than oranges. There are 3 oranges.
BOWL 10:  There are 6 apples. There are 4 fewer oranges than apples.

B. Major statement: *There are oranges and apples in each bowl.*
**Note:** All students need to be familiar with this statement, which is the base for all the following clues.

BOWL 1:   Half of the fruit is oranges. There are 12 pieces of fruit.
BOWL 2:   Twice as many are apples. There are 2 apples.
BOWL 3:   There are twice as many apples as oranges. There are 3 oranges.
BOWL 4:   There are 7 oranges. There are 4 fewer apples.
BOWL 5:   There are 9 pieces of fruit. There are 3 more apples than oranges.
BOWL 6:   There is 1 more apple than there are oranges. There are 9 pieces of fruit.
BOWL 7:   There are 2 oranges for every apple. There are 9 pieces of fruit.
BOWL 8:   There is 1 more apple than there are oranges. There are fewer than 8 pieces of fruit.
BOWL 9:   There are 6 pieces of fruit in all. There are more apples than oranges.
BOWL 10:  There is an odd number of both types of fruit. There are 10 pieces of fruit.

C. Major statement: *There are oranges, apples, and bananas in each bowl.*

**Note:** All students need to be familiar with this statement, which is the base for all the following clues.

BOWL 1: There is an odd number of oranges.
There is an even number of both apples and bananas.
There is a total of 7 pieces of fruit.

BOWL 2: There is an odd number of oranges.
There is an even number of both apples and bananas.
There are 13 pieces of fruit.

BOWL 3: There are twice as many bananas as apples.
There are 2 oranges.
There are 2 more apples than oranges.

BOWL 4: There are half as many bananas as oranges.
There are twice as many bananas as apples.
There are 2 bananas.

BOWL 5: There are half as many bananas as oranges.
There are twice as many bananas as apples.
There are 3 apples.

BOWL 6: There are 2 oranges for every apple.
There are 3 oranges for every banana.
There are fewer than 12 pieces of fruit.

BOWL 7: There are 2 fewer apples than oranges.
There are 3 fewer bananas than apples.
There are 6 apples.

BOWL 8: There are 3 times as many bananas as oranges.
There are 15 pieces of fruit in all.
There is an odd number of apples.

BOWL 9: The number of apples and oranges together is twice that of the bananas.
The total number of fruits is fewer than 10.

BOWL 10: There is an even number of each type of fruit.
There is a total of 12 pieces of fruit.


BIRDS ON A WIRE: CLUE STATEMENTS

A. Major statement: *There are bluebirds and robins on a wire.*

**Note:** All students need to be familiar with this statement, which is the base for all the following clues.

WIRE 1: There are 7 robins.
There is a total of 9 birds.

WIRE 2: There are 2 more robins than bluebirds.
There are 7 birds in all.

WIRE 3: There are 3 fewer bluebirds than robins.
There are 5 bluebirds.

WIRE 4: Every other bird is a robin.
There is a total of 10 birds.

WIRE 5: There are exactly the same number of each kind of bird.
There are a dozen birds.

WIRE 6: There are 12 birds on a wire.
Half of them are robins.

WIRE 7: There are 5 more bluebirds than robins.
There are 4 robins.

WIRE 8: There are 2 bluebirds for every robin.
There are 3 robins.

WIRE 9: There are 2 bluebirds for every 2 robins.
There are 6 robins.

WIRE 10: There are 2 bluebirds for every 3 robins.
There are 6 robins.


B. Major statement: *There are sparrows and bluejays on a wire.*
**Note:** All students need to be familiar with this statement, which is the base for all the following clues.

WIRE 1: There are twice as many sparrows as bluejays.
There are 5 bluejays.

WIRE 2: There are half as many sparrows as bluejays.
There are 12 birds in all.

WIRE 3: For every 1 sparrow there are 3 bluejays.
There are 3 bluejays.

WIRE 4: There are two times as many sparrows as bluejays.
There are 3 bluejays.

WIRE 5: There are two times as many sparrows as bluejays.
There are 8 bluejays.

WIRE 6: There are 2 sparrows for every 3 bluejays.
There are 6 bluejays.

WIRE 7: There are half as many bluejays as sparrows.
There are 8 sparrows.

WIRE 8: There are half as many bluejays as sparrows.
There are 6 bluejays.

WIRE 9: There are 4 sparrows for every 3 bluejays.
There is a total of 14 birds.

WIRE 10: There are half as many sparrows as bluejays.
There are fewer than 15 in all.

C. Major statement: *There are bluejays, robins and sparrows on a wire.*

**Note:** All students need to be familiar with this statement, which is the base for all the following clues.

WIRE 1: There is a total of 14 birds.
There are twice as many sparrows as robins.
There are 2 robins.

WIRE 2: There are half the number of bluejays as robins.
There are 6 sparrows.
There are 2 fewer robins than sparrows.

WIRE 3: There is an even number of each kind of bird.
There is a total of 10 birds.
There are 2 bluejays.

WIRE 4: There are a dozen birds in all.
There are 3 bluejays for each robin.
There are 4 sparrows.

WIRE 5: There are 3 robins.
There is a total of 11 birds.
There are 4 more bluejays than sparrows.

WIRE 6: There are a dozen birds in all.
There are 3 bluejays for each robin.
There are more than 4 sparrows.

WIRE 7: There are 4 robins for each bluejay.
There are 13 birds.
There is more than 1 bluejay.

WIRE 8: For every 1 robin there are 2 sparrows.
For every 2 robins there are 2 bluejays.
There are 2 robins.

WIRE 9: For every 1 robin there are 2 sparrows.
For every 2 sparrows there are 3 bluejays.
There are 12 birds.

WIRE 10: There are 14 birds.
There are more bluejays than sparrows.
There are 5 robins.

. . . . . . . . . . . . .

# 8. PATTERNS WALK

*All ages*

*Content*
geometry, logic, number

*Materials*
your environment

*Why?*
- Provide an understanding and appreciation for mathematical patterns in life.

*How?*

*Step 1:* As a class, go "exploring" for examples of patterns in the world around you. For example: fences, sidewalks, windows, gardens, etc. Students and/or you can carry writing tools to record discoveries; when appropriate, students can collect samples to bring back to class for further discussion and display; rubbings or instant photographs can also be used to record discoveries.

*Step 2:* Once back in class, give students the opportunity to discuss and describe their observations and discoveries — in pairs, groups, whole class, or in individual journals. A patterns bulletin board can be made with drawings and descriptions of the discoveries.

*Thinking It Over*
This activity should be at the beginning of a unit on patterns,

geometry, or even number. Students need to see that the discoveries, discussions, and activities in the classroom relate to what occurs in the world outside the school walls. Starting with this exploration shows them that math is real and that it can be fun.

This activity also works when students are experiencing difficulty with geometry or other geometric concepts. If this happens in your class, take time for a *Patterns Walk*. All of you will be surprised at the mathematical patterns around you.

STUDENT RESOURCE MATERIAL

Carle, Eric. *My Very First Book of Shapes*. Harper Collins, 1985.
Gillham, Bill, and Susan Hulme. *Let's Look for Shapes*. Putnam, 1984.
Hoban, Tana. *Circles, Triangles, and Squares*. Macmillan, 1974.
_____. *Shapes, Shapes, Shapes*. Greenwillow, 1986.
Hutchins, Pat. *Rosie's Walk*. Macmillan, 1971.
Jonas, Ann. *Round Trip*. Scholastic, 1987.

## 9. INVESTIGATING BLOCKS

*All ages*

*Content*
geometry

*Materials*
pattern blocks, geoblocks, tangrams, unifix cubes, and any readily available geometric shapes

*Why?*
• Understand common patterns and geometric shapes.

*How?*

*Step 1:* Begin by asking students to choose from any blocks or geometric pattern pieces that are available in the room. Encourage them to construct larger patterns and creations from the blocks.

*Step 2:* After each building period (15-30 minutes), have each student record their "creation" by drawing the pattern or shape that they have created and writing a description or caption to explain what they have done.

Encourage discussion of what students have tried to do and what they have actually done. Take time to read aloud to the class students' descriptions of their work. Post students' drawings or gather them into booklets so that students can read them during break time or silent reading.

It is important that students understand that the process is open-ended — there is no right answer. Any one problem can be approached from many directions, and may be solved in a variety of ways.

This free exploration can take place over several weeks with students working alone or in small groups.

*Progression 1*

*Step 1:* After a few weeks of free exploration, begin to focus the task by asking students to investigate a specific problem: given a set of blocks or shapes, ask students: "What patterns can you create from these shapes?" or "What larger geometric shapes can you make from these smaller shapes?"

*Step 2:* Have students work with a partner or in small groups. Again, give them plenty of time to investigate and explore, with ample time for a discussion of their findings and creations. Questions such as "What discoveries have you made so far? How have you been able to record your ideas?" can be used to keep students on task.

*Step 3:* If students are more advanced, tie the activity to other issues in math by asking questions such as "What examples of ratio and proportion can you create from these shapes? Can you find ways to demonstrate any rules of fractions, geometry, or arithmetic that we have discussed in class?"

Ensure all students have time to write (draw) about what they have done. At the very least, students should be reading what they have written to the other members of the group. With primary students their drawings will usually constitute their "writing." With encouragement, the simple captions or simple descriptions they provide will expand as they gain experience and confidence.

*Progression 2*

*Step 1:* Limit the range of materials available to students. For example, ask students to use only 2-3 blocks to build something. Then have them describe, in writing, their figure in such a way that other students can build the same figure from the written directions.

*Step 2:* Have students work in pairs or teams to edit and rewrite the directions so that they are clear to others. When students are satisfied with their directions have them put the directions on one piece of paper and the diagram on another. Glue the two pages together, back to back, and distribute the "task cards" to other groups or students who will be required to draw the object described in the directions without peeking at the other side.

## Thinking It Over

As students work through these various progressions you will find that your role will move away from that of a director to one of a facilitator and guide. Your primary role is to motivate students to ask their own questions as they search for answers and novel solutions. Allowing students to explore their own paths of inquiry is as important as anything they discover about geometric shapes or patterns.

When students generate directions for others to follow, they have a meaningful context for the development of their writing skills. Because students are writing to convey meaning, spelling and punctuation should not be an instructional focus. At first, allow students to concentrate on their ideas and creativity. If you plan to publish or duplicate students' work, writing mechanics and "correctness" can be addressed as an activity outside *Investigating Blocks*.

STUDENT RESOURCE MATERIAL

Carle, Eric. *The Very Busy Spider*. Philomel Books, Putnam Publishing Group, 1989.

Emberley, Ed. *Ed Emberley's Big Orange Drawing Book*. Little Brown, 1980.

Ernst, Lee, and Lisa C. Ernst. *The Tangram Magician*. Abrams, 1990.

Hoban, Tana. *Circles, Triangles, and Squares*. Macmillan, 1974.

Kilroy, Sally. *Copycat Drawing Book*. Dutton, 1981.

. . . . . . . . . . . . . . . .

## 10. OVER THE WALL

*All ages*

*Content*
geometry, logic

*Materials*
pattern blocks or geoblocks, pencils, paper, student-erected wall
(book or stiff paper), butcher paper, felt pen

*Why?*
- Develop vocabulary and strengthen general language skills.
- Experience geometry.

*How?*
This activity can be done in several ways depending on students'
writing abilities.

Children who cannot yet write work in pairs. They erect a wall
between them by propping up a book or holding a stiff piece
of paper. Each partner has a set of 3-5 pattern blocks that match.

Partner A — the architect — builds a structure with blocks on
one side of the wall. Partner A then describes the structure to
Partner B who attempts to duplicate it by listening to the clues
and descriptions of Partner A. Neither partner can look over the
wall until both are satisfied that they are finished.

*Progression 1*
To make the activity more difficult, a rule may be given that only
architects can speak and that architects cannot use color when

describing their structures. For example, rather than say "the yellow piece" students must employ the appropriate geometric term such as *hexagon* or *square*. During the activity, circulate and record the nature of the key terms being employed by students.

*Progression 2*

Place students who can write in groups of three. One team member is the architect, one the builder, and one the recorder. The process already described still holds except that the recorder writes down all words used that are mathematical or directive in nature such as *same, on top of, rectangle, side*. The use of geoblocks for these students stimulates greater vocabulary development and discussion.

When each student has had at least one opportunity at each role, it is time to process the activity. With younger children you can brainstorm a list of words they used, or read what you (or an aid) recorded as you observed the activity. With students who created their own lists, transfer these words to master lists for discussion. Be sure to list these words on paper so they can be hung around the room for future reference.

*Thinking It Over*

Students often find that as they work to describe their structures they come across the need for a word or phrase they don't yet have in their vocabulary. This is the moment when they are open to learning. Rather than introduce vocabulary through a list on the board or on a worksheet, *Over the Wall* and similar activities (see *Grab Bag*) present a more natural situation for learning. Math vocabulary is learned in an active, meaningful, social context.

STUDENT RESOURCE MATERIAL

Brown, David. *The Random House Book of How Things Were Built.* Random House, 1992.

Dunham, Meredith. *Shapes: How Do You Say it?* Lothrop, Lee & Shepard, 1987.

Macaulay, David. *Castle.* Houghton Mifflin, 1982.

————. *Cathedral.* Houghton Mifflin, 1981.

————. *Pyramid.* Houghton Mifflin, 1982.

. . . . . . . . . . . . . .

# 13. MEASUREMENT TOOLS

*Ages 8–12*

*Content*
measurement, number, logic

*Materials*
string, cubes, grid paper of various sizes, stop watch, measuring cups, rulers (metric and standard), yard sticks, metresticks, scales, thermometers, paper cups, toothpicks, or whatever else is available in the classroom

*Why?*
- Understand that different measurement tools are required depending upon the size and use of what is to be measured.
- Understand the use and value of standard and non-standard units of measure.

*How?*
The purpose of this activity to introduce students to the fact that objects can be measured in different ways depending on the "standard" of measure adopted.

*Step 1:* Generate a list of items around school that can be measured. You may wish to develop the list yourself or have students brainstorm items. Such a list might include: the playground, a pencil, fingernail, the room, a desk, football field, flagpole, jars, a ball, globe, etc. Choose one object to measure, for example, the length of a table. Have students measure it using a variety of non-standard measurement units:

toothpick, a paper clip, or a pencil length.

*Step 2:* When the measuring is completed, have students compare their findings. Ask questions similar to the following: "How many pencil lengths did it take? How many toothpicks? Is everyone's answer the same? Why not? Should we use the same units of measure to measure the football field? If so, why? If not, why not? Will we always get the same answer?"

## Progression 1

*Step 1:* After this discussion, have students measure the length of the playing field by pacing it off. Again, is everyone's answer the same? If not, why not? When is this type of estimation good enough? When is it not?

*Step 2:* Be sure that during and after completing these activities and discussing them, students make entries in their logs or journals. They can, for example, keep track of their processes, observations, and revelations. Allow them time to illustrate their logs or journals and add diagrams, if appropriate.

*Step 3:* Discuss the purpose and value of standard measures using questions similar to the following: "Do the grounds keepers in professional sports simply pace off the field? Is an estimation of the size of a desk good enough when buying it at the store? Is it good enough simply to estimate the size of the carpet for the classroom? When is close enough good enough? When is it not?"

## Progression 2

As a follow up, explore various ways different things can be measured. For example: A desk (height, weight (mass), volume, construction time, cost, age). Why and when are these ways of measuring important? Examine measurement variations with several listed items.

## Progression 3

Have students select other objects not discussed or listed and have them write about how to measure these objects in different ways. Such writing can also focus upon how and why these different measurement units are or are not useful.

*Thinking It Over*

Although this activity may seem simple, the concepts associated with units of measures and standards are quite complex. How many adults do you kow who claim that they just can't seem to understand the metric system? Giving students an experiential understanding of different units and standards of measures is an important first step toward facility with the many standards of measurement employed today.

STUDENT RESOURCE MATERIAL

Ardley, Neil. *Making Metric Measurements.* Franklin Watts, 1984.
Laithwaite, Eric. *Size: The Measure of Things.* Franklin Watts, 1988.
Pluckrose, Henry. *Length.* Franklin Watts, 1988.
Shapp, Martha, and Charles Shapp. *Let's Find Out About What's Big and What's Small.* Franklin Watts, 1975.
———. *Let's Find Out About What's Light and What's Heavy.* Franklin Watts, 1975.

. . . . . . . . . . . . . .

## 14. BODY MEASUREMENTS

*All ages*

*Content*
measurement, number

*Materials*
string, scissors, tape, journals

*Why?*
- Understand that measurement is a process of comparison.
- Understand that various everyday measurements often have relationships involving ratio and proportion.

*How?*

*Step 1:* Each student chooses a body part and measures its length or circumference with a piece of string. This personal measure becomes the standard measure for the rest of the activity.

*Step 2:* Now students can compare the length of string with other body measurements: leg, wrist, waist, nose, etc. How does it compare? Help younger students frame their answers to this question using fractions and other measurement terms. For example: "The circumference of my head is half as long as the length of my leg." If necessary, help students fold their string to measure smaller lengths; show them the number of times their length of string is needed for a particular measurement.

*Step 3:* Each student can develop a "Body Book." Students draw and write the base-line measurement and the comparison measures they have taken.

*Progression 1*

Have students choose a different body measurement as their standard of measure, and repeat the process. Students may also wish to compare their body measurements with other things: tables, doors, floor and, of course, other students. As in the first activity, encourage students to express their findings using terms related to ratio and proportion.

*Progression 2*

This activity can be very effective in conjunction with a unit on the history of measurement — its uses and evolution beginning with ancient civilizations. Digits, spans, paces, cubits, yards, and other units of measure developed before the metric system were based on relationships to the human body. When introduced in this manner, students might imagine themselves as Egyptian farmers or pyramid builders, adding a different perspective to the activity.

*Progression 3*

Have students measure their height using one of their feet as a standard measure. Have them answer the question: How many of your "feet" tall are you? Have students compare their heights with those of their friends and classmates. Are their ratios the same? Why? Why not? How do these ratios compare with adults? With babies? Charge students with finding out. This makes for an excellent homework assignment.

One way to illustrate the results of this investigation is for each student to trace the outline of one of their feet and make several copies. In pairs, students measure each other's height against the wall or on the floor using the "foot" (the standard measure) of the student being measured. Students tape copies of these "feet" up the wall to a point even with their heights. A classroom decorated with all the students' "feet" walking up a wall illustrates a point about body ratios as well as making a very interesting display.

*Thinking It Over*

The process of comparing two things for length marks the beginning of facility with measurement. When children first use their own bodies as measurement tools, they will more readily understand measurement concepts. After much experimenting, students come to understand that establishing standard units to which different objects can be compared simplifies and clarifies the measurement process.

STUDENT RESOURCE MATERIAL

Briggs, Raymond. *Jim and the Beanstalk*. Penguin, 1973.

Greenfield, Eloise. *Big Friend, Little Friend*. Writers and Readers, 1991.

Myller, Rolf. *How Big Is a Foot?* Dell, 1991.

. . . . . . . . . . . . . .

## 15. BODY SCALE

*Ages 8–12*

*Content*
measurement, number, probability and statistics

*Materials*
butcher paper, pencils, crayons, string, staplers, metresticks and other measurement tools, journals

*Why?*
• Develop an understanding of ratio and proportion.

*How?*

*Step 1:* Tell students they need to create a butcher paper representation of their own bodies at a scale of half their size. Introduce the assignment in general terms, that is, without providing explicit, step-by-step directions to students.

*Step 2:* Place students into groups so that they can discuss how to accomplish the task. It is very important that students have the opportunity, among themselves, to wrestle with the requirements of the task. If the groups have difficulty generating questions or conceptualizing the problem, ask: "What measurements do you need to make?" "Your paper is flat but your body isn't; so how will you show your waist, head, feet?"

Part of the responsibility of the groups is to keep a written record of the steps they take in completing the task, the problems they face, and any observations they make.

*Step 3:* By this stage there will be a room full of half-size replicas. Bring the groups together so the whole class can discuss how they developed the representations of themselves. Ask: "Did you have a plan or strategy for reducing your figure?" "What did you do?" "Is there a pattern or formula that you used?" "Can you explain ratio and proportion?" "What problems did you have?" "How did you solve them?" It is from this final discussion that you can draw out the principles at issue in the concepts of ratio and proportion.

*Thinking It Over*

It is important to let students discover their own process for completing the assignment. As their teacher, help them explore the process and try not to worry too much about the product. Take lots of time throughout the activity to stop and discuss what troubles and solutions they are experiencing.

Successful completion of the activity requires that students deal with the issues of proportion and scale: a very difficult concept for most of us to grasp well. By starting with their own bodies, students will have a personal, real-world referent to draw upon when they later work with ratio and proportion in more abstract ways.

STUDENT RESOURCE MATERIAL

Briggs, Raymond. *Jim and the Beanstalk*. Penguin, 1973.
Carroll, Lewis. *Alice's Adventures in Wonderland*. Various editions.
Myller, Rolf. *How Big Is a Foot?* Dell, 1991.
Lord, John and Janet Burroway. *The Giant Jam Sandwich*. Houghton Mifflin, 1987.
Swift, Jonathan. *Gulliver's Travels*. Various editions.

## 16. C L A S S R O O M   M O D E L

*Ages 8–12*

*Content*
measurement, number, geometry, probability and statistics, logic

*Materials*
paper, pencils, string, rulers, metresticks and other measurement tools, centimetre grid paper (on card stock, if possible), scissors, glue, tape, journals

*Why?*
- Develop an understanding of ratio and proportion and their relationship to various measurement tools.
- Develop a spatial sense and appreciation for structural design.

*How?*
Introduce this activity through a discussion of the design and construction of buildings. Specifically, this discussion should address the need for and purpose of blueprints and standard measures. Discussion of these larger issues provides a context for the activity.

*Step 1:* After the initial discussion, teams of students discuss the process for building a scale model of their classroom. As described in *Body Scale*, it is important that students not be given detailed directions but rather that they wrestle with the problem on their own.

*Step 2:* Supply students with centimetre grid paper and various measurement tools. After student teams discuss and develop a plan, but before they actually start to measure or build, have them individually record in their journals their joint plan of action.

*Step 3:* Next, still in teams, students measure the classroom, using whatever tools they decide upon. Based on these measurements, students create their models. Expect the models to run the gamut from crudely simple to ornately complex.

*Step 4:* After models have been completed, be sure to focus discussion on the differences among the various teams' approaches to the task, not on whose team created the "best" model. The objective in this activity is to show that ideas and discovery are more important than the product. To highlight the process, ask questions such as: "What were the common problems encountered?" "How were they solved?" Attend to students' creative endeavors and different approaches to the problem.

In addition to questions regarding process, take time to explore with them the specific math skills necessary for completion of the problem. Students are more attentive to skills instruction when they have an immediate need for a skill.

*Step 5:* Students create a written record (including drawings) of their discoveries in their journals. Such documentation should be an ongoing process and final step.

*Thinking It Over*

Creating classroom models through team work provides opportunities for learning in a variety of dimensions. In addition to the application of measurement and other math skills, students must interact, set directions and plans, justify their decisions to others, as well as work cooperatively. Throughout all of this students will be talking to one another using the language of mathematics.

The writing that students generate is also very important. It helps students formulate and organize their thoughts. It provides them with a record of their thinking and progress. It also serves as an excellent assessment tool for you. Were students able to write about the problems and solutions their team discovered? How did they describe and interpret the process? Did they discover the need for and purposes of standard measure and scale?

Macaulay, David. *Castle*. Houghton Mifflin, 1982.

_____. *Cathedral*. Houghton Mifflin, 1981.

_____. *Pyramid*. Houghton Mifflin, 1982.

Hinchcliffe, Jo. *The Hilton Hen House*. Ashton Scholastic, 1987.

. . . . . . . . . . . . . .

## 17. TIME LINES

*Ages 6–12*

*Content*
measurement, logic, number

*Materials*
paper, pencils, colored markers, crayons, sentence strips, butcher paper

*Why?*
- Develop a sense of time measurement.
- Help "socialize" students to the clock and its relationship to themselves.

*How?*
A variety of time lines can be drawn by students to chronicle their life experiences and help them develop a better understanding of the conventions our culture has developed for marking time.

The core activity for *Time Lines* is the creation of illustrations and/or written descriptions that chronicle events as they occur over a given time in students' lives.

*Step 1:* Begin with a class or group discussion about a typical day: "What time do you get up in the morning? What things to you do before you go to school? In what order? At what time?"

Decide how much time you want to spend. This part of the lesson can be, if you wish, a long brainstorming session. With students, develop lists of activities by the hour and/or by the specific day of the week. Conclude the discussion by choosing a set period of time to chronicle, such as a typical Friday. After students have several events on the list: getting out of bed, morning recess, silent reading, feeding the dog, arguing with my sister, homework ... decide on a limited number of scenes for this first time line, say: waking up; arriving at school; lunch; mathematics activites; after school; dinner.

Next, discuss what times should be associated with each event. Lunch should be fairly simple. Waking up, after school, and dinner may involve more discussion.

*Step 2:* Once the time frame has been set, have students — either individually or in small groups — create a pictorial/written story sequence of the events that occurred during the time in question.

For example, young primary students may be guided to separate scenes for their morning, noon, and night; breakfast, lunch, and dinner; or scenes of their own choosing.

Don't make all the decisions about the exact times for each event. Now is the time to turn the process over to small groups. Have each group create their own time line. They need to decide what time goes with each event, and then record it as a clock face and/or digital display somewhere on each of the illustrations they create.

*Step 3:* Have students choose a format for their time line:

1) Use large butcher paper and create a mural with the sequence of events and captions or descriptions either written balloon-style in the illustrations, or written on lined paper and then glued under each scene.

2) Use individual worksheets for each event. Create booklets or fix the events together in sequence to hang about the room. For fun, have groups hang their scenes with clothes pins. After they have described their sequence, move a couple of scenes around and ask them to tell a new story as dictated by the change in order.

3) Have groups use sentence strips or larger rolls of paper to create a scroll of events.

*Progression 1*

Opening Up: After your students' first attempt, let them try "A Typical Saturday." Decide on a limited number of events — four, six or eight — only don't name what they are to be. Let the groups decide what scenes belong in a typical Saturday.

Other days they might develop time lines for: field trips; vacations; birthdays; the most important day of my life, the day I was born. The work can be orally presented to the class, posted, or published.

### Other Progressions

1) Weeks. One illustration for each day or as few or as many as students feel are needed to chronicle a given week.

2) Seasons; summer; summer vacation; the entire year with an emphasis on nature's cycle. Time lines can be connected into a circle rather than as lines to reinforce the cyclical nature of the seasons or other events. Illustrations can be on both sides, or descriptions on one side and representations on the other.

3) Family Histories: individuals, immediate family, or through the generations.

4) The future: This year, next year, the next 10 years, 50 years.

5) Brainstorm!

### Thinking It Over

Until students are eight or nine years of age (often older) they may not have the logical thought processes to understand time as we use and measure it in our culture — from seconds, minutes … centuries, millennia, and beyond. The activities suggested here attempt to personalize the concept of temporal sequencing while reinforcing students' ability to tell time by the clock or the calendar.

STUDENT RESOURCE MATERIAL

Hutchins, Pat. *Clocks and More Clocks*. Puffin Books, Penguin, 1974.

Anno, Mitsumasa. *All in a Day*. Putnam, 1990.

Carle, Eric. *The Very Hungry Caterpillar*. Philomel Book, Putnam, 1989.

. . . . . . . . . . . . . .

## 18. C A L E N D A R   D A Z E

*All ages*

*Content*
measurement, number, logic

*Materials*
paper, pencils, colored markers, crayolas, calendar template (see example at the end of the activity)

*Why?*
- Develop a sense of patterning: sequence, before and after, direction, graphing.
- Learn to use a calendar.
- Develop an awareness of personal, social, and historical events.

*How?*
The calendar is a powerful teaching tool. It provides an opportunity to integrate teaching in mathematics, history, literature, and almost every other content area. One beauty of the calendar as a teaching tool is that it provides a functional and meaningful real-world experience for students.

Whichever way you choose to use the calendar in instruction, be sure to encourage the students' imaginations and creativity. Allow lots of time for discussion and writing. Individuals and groups should personalize their activity as much as possible.

*Step 1:* No matter what the age level, an excellent opening activity before any extensive discussion or instruction about the calendar is an assessment of what students already know.

Keep the instructions simple. Explain that everyone uses calendars every day and in many ways. Ask the students to make their own calendars. Then have students write as much as they can about how calendars are used and whatever else they know about them. Primary age children can draw their impressions and/or dictate their ideas to an older student or the teacher.

Allow plenty of time, at least one class period. This should not be given as a homework assignment because it is important that they do all the work by themselves. This class activity helps you assess where students are at the beginning, and provides you and them with a measure of how far they have come when these activities are completed.

*Step 2:* After students have designed and written about the calendar as they know it, work as a whole class or in small groups to brainstorm about the calendar: What is it used for? How do you or your family use a calendar? What are some words we use with the calendar? Where do these words come from?

Generate lists on larger pieces of paper and hang them around the room so that as students work through more activities they can refer to the list and remind themselves that they already knew a lot before they started. Add to the lists as students learn more about the calendar.

*Progression 1: This Month*

Use the calendar template provided here so that each student can fill in the details of the current month. If necessary, fill in the days of the week for them before you reproduce it for their use.

*Step 1:* When required, help students fill in the first and last day of the month. Continue with the dates for the rest of the month.

*Step 2:* Consider the month as a broken number line. You and/or your students take a template that has been filled in and cut out each week as a separate line of dates. Next, arrange them so that they are all in a sequential line to help students understand the idea of a number line broken in a pattern of sevens (this same idea is explored in the Hundred Chart activities, p. 00).

As a variation, cut the month into irregular puzzle pieces that students align to piece back together. Have students create such puzzles for their classmates to solve.

*Step 3:* Explore the whole month. Find today's date, then ask students to identify: *the day after tomorrow; the day before yesterday;* and *next Sunday*, etc. Find other dates of interest: class birthdays; birthdays of famous people; holidays; upcoming school and community events, etc.

*Step 4:* Work with number patterns. Ask students: "What day will it be in ten days? What day will it be in two weeks? Have students make up their own questions for one another.

*Progression 2: Days of the Week*

Explore the origin of the names of the days of the week, the names of the month. Ask: "Why are these the names of the days? Does anyone know where these names come from?" Have volunteers research the origin of the names of the days and months. Also ask if anyone knows poems, rhymes or songs about the days or months. Students can also make up their own names and origins for the names of the days and months.

*Thinking It Over*

Calendar activities provide a wide range of instructional opportunities. Calendars can simply be used to become familiar with dates and numbers. The concept of time and how we organize it in our culture is quite confusing for students. Take it slowly. Recognize developmental differences in your children. Spread these activities out over the entire school year.

If you do go beyond simply using calendars to work with dates and numbers you'll find yourself studying agricultural practices, religion, anthropology, botany, and biology just to name a few. Not to worry. Let the class take you as far as they want to go. This can be a truly open-ended activity.

STUDENT RESOURCE MATERIAL

Carle, Eric. *The Very Hungry Caterpillar*. Philomel Books, Putnam, 1989.
Pluckrose, Henry. *Time*. Franklin Watts, 1988.

**Note:** *For student use, enlarge the activity sheet by 130-150%.*

CALENDAR DAZE

. . . . . . . . . . . . . .

## 19. MEASUREMENT

## INVESTIGATIONS

*Ages 8–12*

*Content*
measurement, number, logic

*Materials*
pencil, paper

*Why?*
- Develop a sense of how and why things are measured in the way that they are.
- Develop skills of inquiry and investigation.

*How?*

*Step 1:* Begin the activity with a discussion by brainstorming answers to the question: ''What things to do we measure?'' Answers might include: the length of a football field, our body's weight (mass), hours in a school day, speed of a car, amount of water used in a day, milk in a carton. This list can, and should, be very long. Further, it should be open-ended and available for the addition of other items as they arise — after lunch, tomorrow, or next week.

*Step 2:* Next, decide upon a fixed list of objects from your list and then ask: ''How do we measure each of these items?'' Move to sorting the list into categories. Let the students determine the categories. They may work toward groupings such as time, length, weight (mass), volume, etc.

*Step 3:* Choose one or a few items that were brought up during brainstorming. Expand upon them to create interesting measurement investigations. For example, the amount of water used during a day by your class, the school, the town; cubic metres in a room; weight (mass) of all the students in the class; number of heartbeats in the classroom in a minute, an hour, a day; or the distance a pet dog, cat, or hamster walks in a day, a week or a month. Always encourage the addition of ideas to the original list.

Once the list has been developed, break students into groups and either assign or have students choose one of the items. Each group's assignment is to design a measurement system or strategy for the item that they have been assigned. Initially, you may have to help the groups get going by modeling key questions that have to be answered.

This is an open-ended task. The group may design tools, simply work out the numbers involved, or make inquiry phone calls to gather information to solve the task. The exact nature of what they will need to do depends on the topic they have chosen.

As with all activities, when their investigations are completed, students should share and discuss their findings. Group presentations to the class are varied and fun.

*Thinking It Over*

*Measurement Investigations* helps students to understand the relationship between situations in the real world and measurement systems and techniques. Rather than have students learn measurement skills in isolation and then apply that knowledge to the world, this activity allows students to understand how measurement tasks serve specific functions in our day-to-day lives. The skills arise from a specific need.

Brainstorming is important to this activity. Be sure to accept all suggestions without judgment, as long as they aren't too disruptive. The idea is to stretch the imagination as well as to encourage thinking and participation.

STUDENT RESOURCE MATERIAL

Blocksma, Mary. *Reading the Numbers: A Survival Guide to the Measurements, Numbers and Sizes Encountered in Everyday Life.* Penguin, 1989.

. . . . . . . . . . . . . .

# 20. NUMBERS ALL AROUND US

*All ages*

*Content*
number, geometry

*Materials*
butcher paper or chalkboard, pencil, paper

*Why?*
- Develop concept of number.
- Link concept of number with cultural and functional literacy.

*How?*
*Step 1:* Create a chart on the board or butcher paper. Butcher paper is better because the chart can be extended. Make 12 columns (or as many as you think appropriate — 12 seems to work well). Mark each column at the top with a number 1 through 12.

The goal of this exercise is to list things in the world which come in or occur in that number. Accept responses that use the number in a common expression or other usage. (Be flexible in this regard; accept a broad range of responses.)
**Examples**
1: One head, eye in a cyclops, wheel on a unicycle, nose on a face, god in many religions.
2: Legs on a human, eyes in a head, wheels on a bicycle, parents in some families, parts to the yin and yang.
3: Sides on a triangle, parts of the Trinity, faces of Eve, etc.

*Step 2:* Solicit a few responses for some or all of the categories. Then have the students work in groups to expand the columns on their own group chart. After some time, ask for additions to the chart. Fill in the big chart. Expect a lively discussion.

*Step 3:* This activity can go on for quite a while, so find a comfortable stopping point, but tell students you know they will think of more things to add. Take a few minutes each day to ask if there are any additions to the chart. You will find many students have taken the discussion out to the playground and home to their families.

### Progression 1

Use this chart to initiate discussion about addition, multiplication, and division. If, for example, "legs on a dog" is listed in the 4 column on your chart, ask: "If we had two dogs, how many legs would there be? What if we had four dogs?"

The number chart and the subsequent discussion provides an opportunity to link addition and multiplication. The columns and entries in this activity utilize objects and concepts from the child's world as a vehicle for arithmetic instruction.

### Thinking It Over

Mathematics has its foundation in real-world situations. The concept of number must first be tied to a learner's experiences. This activity makes that connection for the individual while drawing upon the whole class to build a larger bank of references.

A byproduct of this activity is expanded multicultural literacy. One child may add "Five players on a basketball team," or "six sides to a honeycomb," or "seven (nine) candles on a menorah." This may be new information for some students in the class. *Numbers All Around Us* provides a way for students to introduce their individual cultures or interests to others in the class: everyone learns.

STUDENT RESOURCE MATERIAL

Aker, Suzanne. *What Comes In 2's, 3's, & 4's?* Simon & Schuster, 1992.

Crews, Donald. *Ten Black Dots.* Greenwillow, 1986.

. . . . . . . . . . . . . . .

## 21. E S T I M A T I O N

*All ages*

*Content*
measurement, logic, number

*Materials*
objects, large and small, in clear containers found in the home
and the classroom; catalogs; shopping circulars

*Why?*
- Establish an understanding of the importance and utility of
  estimation skills.

*How?*
*Step 1:* Introduce the concept of estimation with questions similar to the following: "Which things are easy to count and which
things are more difficult? Why?" (Choose examples from the
students' immediate environment to anchor the discussion.)
"Which is easier to count, the number of chairs at your table
or the number of chairs in the room? The number of windows
in the room or the number of windows in the school? The number of cubes in this jar or the number of beans in this jar?"
Probe students' responses. Ask: "Why do you think so?"

*Step 2:* After the discussion, ask students: "What are some different ways or strategies for finding the answers to these questions? What are some ways we could find out the number of
windows in the school, or beans in the jar?" List the various
methods students suggest on the board, or have them work

in groups first to generate ideas. You may want to have them draw or write about their ideas.

*Step 3:* Once various approaches to answering the questions have been offered, ask students: "When is it important to know an *exact* answer and when is *close enough* O.K.? Why?" Again, use examples from students' immediate environment to anchor the discussion using questions similar to the following: "Is it important to know the exact number of lunches ordered for the class? If so, why? If not, why not?" Use the same approach with other questions such as: "How many kids are on the playground? How many days are left until school is out? How much time is left until lunch? What is the cost of your favorite food?"

Students can cooperatively generate lists of questions that require exact answers and those that do not require exact answers. Ensure students understand the question: "When is estimation good enough?" Allow ample time for them to discuss different strategies for estimation before they generate their two lists.

## Progression 1: Estimation Jars

Fill clear containers with different objects: candies, macaroni, toothpicks, marshmallows, beans, blocks, or whatever is available and will fit. The number of marshmallows in a jar is easier to guess than the number of toothpicks. Start with easy objects!

Each day of the week have at least one estimation jar set out and have each student write out his or her guess and name on a slip of paper and drop it in a "guesses" jar. Make a ritual out of announcing the exact number and which students made estimations closest to the correct number. Be sure to recognize the range of guesses and discuss why some objects produce greater spreads than others. A jar of 27 marshmallows may have an estimation range from 15 to 35. Estimations for a jar of 520 beans may range from 100 to 3000 depending on students' experience with estimating.

As time goes on, continue to change objects and jars. You'll be surprised at how much better everyone's estimation skills become over time, including yours, for you too can play! Don't count the objects before you fill the container. As an alternative, let students bring in containers they have filled. When it's time to reveal the answer, have a group of students count the objects.

Discuss the group's strategy for counting the objects. Some groups, for example, will count by 2's, some by 10's, others by making piles of 50's. There is no one correct strategy for counting.

Many people develop different strategies for guessing. Encourage class discussion about the various approaches focusing on the fact that there is no one right way to estimate.

### Progression 2: Number Estimation

List a series of numbers on the board. Allow students a few moments to look at the numbers, but not long enough to add them in their heads or with a pencil. Encourage estimation of the sum. Any answer should be accepted. After several guesses, have students work through the correct answer. When the correct answer is available discuss the process, eliciting from students the strategy they employed and its relative effectiveness. Ask, too, when this type of estimation is useful. Why?

Multiplication and division problems can be used equally effectively for estimation.

### Progression 3: Using catalogs or shopping circulars

Have students quickly flip through the pages and pick a number of products they would like to purchase. Ask first for rough estimates of the total cost, then have them give the exact answers with pencils or calculators.

### Progression 4: Have a party!

Make-believe that the class is having a party. Ask students to generate a list of what will be needed. Write the list on the board. Decide how many of the items will be needed, how they are packaged, and what they might cost. For example:

|  |  | Cost per item | No. required |
|---|---|---|---|
| Chips | Bag | $1.89 | 5 |
| Soda | Six pack | 2.49 | 3 |
| Pizza | Box | 7.99 | 4 |
| Alka Seltzer | Packets | 1.25 | 20 |

Have students estimate the total cost of the party. After the initial overall guess, estimation can be done item by item. This activity also provides a great opportunity for students to use calculators with a memory key to find the exact answer.

As a follow-up to the party, encourage students to go on the next family shopping trip and estimate the total cost before the items are rung up at the register.

*Thinking It Over*

Students need practice and experience determining what is meant by the terms *more than, less than, as much as,* etc. They also need experience with not always having a "right" answer. Estimation activities can help students achieve both of these goals. At first, students may be timid about guessing. For any activity based on estimation, it is important that you "play down" the right answer. Praise any attempt. This helps students investigate a variety of strategies and they will soon see that there is nothing to lose by estimating and that nobody is right all of the time, or needs to be right all of the time.

Estimation is a necessary skill in all areas of mathematics. Students should form the habit of estimating first before they use other methods to compute the exact answer. In arithmetic, especially, this cannot be emphasized enough. For example, when students use calculators they should have a good estimate of what an appropriate answer might be. Too often, many of us accept an answer on a calculator's display that is wildly incorrect but we take the answer as correct simply because that's what came up on the screen. It is important to be able to predict with a certain amount of accuracy what outcome is expected.

There are many times in daily life — occupations such as those in financial institutions excepted — when "close enough" is good enough. One of the only places, it seems, where exact answers are expected most of the time is in school. This does not represent a real-world situation, yet students learn early that there is only one "right" answer. Often this serves to limit their thinking and creativity. Yes, kids all need to know their arithmetic skills, and estimation is critical among them. They will develop this essential skill only when it is encouraged.

STUDENT RESOURCE MATERIAL

Asch, Frank. *Popcorn.* Parents' Press, 1979.
Pittman, Helena. *A Grain of Rice.* Hastings House, 1986.

## 22. NUMBER POSTERS

*Ages 7–12*

*Content*
number

*Materials*
paper, pencils, pens, magazines, glue, scissors, whatever else is available

*Why?*
- Review and reinforce the concept of number.
- Reinforce vocabulary related to number.
- Reinforce the connections between the real world and number.
- Reinforce arithmetic skills.

*How?*
Students choose a number between one and one hundred. Their assignment is to make a poster showing, in as many ways as possible, representations of that number. Set a target goal of at least 15 different representations for younger students, and at least 25 representations for older students.

The selected number should be large and visible in the center of the poster, so that when it is displayed it will be clear what number was chosen. For example, a poster of the number 12 would have this number displayed in large, colorful numerals. Written onto the poster might be several arithmetic problems such as $12 \times 1 = 12$, $144 \div 12 = 12$, $6 + 6 = 12$, and so on. Students might also include a photo of the face of a clock cut from a magazine, a dozen eggs drawn and colored or even an egg carton attached, pictures of 2 six packs of soda, or a calendar showing twelve months. The possibilities are limited only by students' imaginations.

As with all activities, allow ample time for discussion and the trading of ideas. You may want to limit the number of students who make posters on the same number. This activity makes for great homework because the whole family will have lots of suggestions.

When the posters are finished and displayed, they become great discussion pieces as well as exciting and meaningful decorations for the room.

*Thinking It Over*

This activity is similar to *Numbers All Around Us* in that it focuses on how numbers are an integral part of our world. This activity helps students view the world from a mathematical perspective. In addition, *Number Posters* allow students the opportunity to review basic functions in arithmetic. An easy way to represent 24, for example, is as $23 + 1$ or $24 + 0$. Students' posters can reflect what they are learning about all areas of arithmetic.

Making these posters is fun, and no two are alike. Be sure to focus on *how different they all are*, because all students bring their own knowledge and perspective to their personal work.

STUDENT RESOURCE MATERIAL

Aker, Suzanne. *What Comes In 2's, 3's & 4's?* Simon & Schuster, 1992.
Crews, Donald. *Ten Black Dots*. Greenwillow, 1986.
Pluckrose, Henry. *Numbers*. Franklin Watts, 1988.
Schwartz, David. *How Much Is a Million?* Scholastic, 1987.

to match those on your chart. These squares can be used as markers on your overhead: they shade the target numbers but let students see which numbers are covered.

Start by asking: "Do you see any patterns in this chart?" For example, students may notice that a certain column of numbers are all multiples of ten. Be sure to take plenty of time for this discussion. You may want to have students work in groups and record their ideas. This can work as an informal assessment of where they are at the beginning of these activities. Later, you and students can read these initial assessments to see how far they've come.

*Progression 1: What's My Number?*

Have students use a separate piece of paper to cover up all but the first row of numbers 1-10. Choose a number between one and ten and play a version of *Twenty Questions*. Begin by stating: "I am thinking of a number between 1 and 10. Does anyone know, for sure, what my number is? Any guesses? I'll give you clues and see if you can determine what my number is." As you provide clues to help students determine your number and students offer guesses, be sure to ask: "What do we know now? Do you know the number for sure? Why or why not? Do you need more information?"

For example, if your number is 6, provide clues such as: "My number is greater than three. My number is an even number. My number is less than seven."

As students attempt to solve the riddle, have them use markers to cover up the numbers they have determined are not the answer. Be sure to prompt students after each clue with questions such as: "Now what do we know?" You can be modeling this on the overhead as students work on their own charts.

This activity allows you to reiterate or reinforce many number concepts that you've covered with your students. For example, you might stress concepts such as *greater than* and *less than*: "My number is greater (or less) than: the number of fingers on one hand; the letters in Tori's name; the number of days in a week; the number of legs on a chicken; 3 + 3; 9 + 4."

Depending on the level of your students' development and skill, continue this activity with larger sections of the chart (see *What's My Number?* in *Cooperative Logic*). As you expand the activity to include all of the numbers of the Hundred Chart, the range of the possible clues increases. For example, such clues might include: "My number is a multiple of 7; a prime number; divisible by 12; a square; greater than the number of states in the U.S.A.; less than my age; greater than my I.Q."

*Progression 2: Patterns*

Begin this activity using the Hundred Chart with the statement: "Mark all the even numbers on your chart beginning with 2." This request can also be stated: "Mark every number counting by twos." Which path you choose depends on the language and conceptual abilities of your students.

Students may either mark the patterns they explore with markers, so that they can reuse individual charts, or if you have the resources, you can duplicate enough charts so that individuals or groups can use crayons or colored pens to create permanent, colorful records of the patterns they discover.

Once students have completed this task, ask: "What pattern or patterns do you see? Describe it/them." Spend time discussing students' perceptions of the patterns created. The idea is to have them view the concept (in this case multiples of 2) in a new way — as a regular, visual, and consistent pattern.

Students can now move on to looking at the patterns created by multiples of three. Before actually coloring or marking the chart, ask students: "What do you think the patterns will look like? Can you describe the pattern?"

After discussing what the pattern might look like, have students mark every third number on their chart. Again, there are several ways to ask them to do this. You can ask them to count by 3's, or work in multiples of 3's.

When students have colored in the patterns of various multiple numbers on their papers, post them around the room so that comparisons can be more easily made, or create pattern books that include written descriptions.

*Progression 3: Multiple Multiples*

In this progression, have students work with more than one multiple simultaneously on a given chart. For example, have students draw a green circle around all the multiples of 3. Then have them put a red diagonal through all the multiples of 4, and a blue cross on the multiples of 5.

Once students have completed this task, ask them: "What do

you notice?" "Why do some numbers have more than one mark?" "What can you tell me about these numbers?" "What can you tell me about the numbers that have only one mark?" "What new patterns do you see?"

This activity provides an opportunity for students to see that there are overlapping relationships among the numbers — that is, that a number can be a multiple of three and five, for example. They are exploring factors.

*Progression 4: Target Numbers*

Choose a number, say 24, on the Hundred Chart to start this activity. Have students place a marker on the "home" or "target" number. Have students add 5 to this number, and then ask: "What number are we on now?" As with the earlier activities, be sure to model the process for students, and allow them to use manipulatives or markers to find the answer. After you have heard from the first volunteer, be sure to ask: "Is there anyone who found a different way to reach their target?" Encourage variety and creativity of strategies.

For many students placing an addition problem in the context of a story is a big help. For example: "There were 23 fleas on my dog. 5 more hopped on. How many fleas does my dog have now?"

Once you all feel comfortable, try some more complicated stories. Be sure to give your students time to follow each statement on their chart before you give them the next bit of the story. Here's an example: I had 52 stamps in my collection. I gave 24 to my friend Lee. My dad brought 12 home from work for me. I bought 36 new ones at the post office. I made a trade with my sister. I gave her 17 in exchange for 5 that I really liked. How many stamps are in my collection now?

As you work to solve the problem presented in words, be sure to work slowly so that students can move their marker on their chart with each new piece of information. When students fully understand the process, have them write their own target-number stories for their classmates to follow.

*Thinking It Over*

Working with Hundred Charts helps students to find other ways of recognizing patterns in the numbers and functions of arithmetic. For example, coloring the multiples of four is another way for children to connect with the concept of multiplication and establish a reference for future work with arithmetic skills. A Hundred Chart is a manipulative that integrates language, concepts, and application of mathematics.

The Hundred Chart also provides another vehicle for the exploration of word problems (see *Number Plays*, *Headlines*, and *Cooperative Logic*). In each case a manipulative or concrete object helps students to understand and solve the problem.

STUDENT RESOURCE MATERIAL

Feelings, Muriel. *Moja Means One: A Swahili Counting Book.* Pied Piper Press, 1976.

Rosenberg, Amye. *1 to 100 Busy Counting Book.* Western Publishing, 1988.

Sloat, Teri. *From One to One Hundred.* Dutton Children's Books, 1991.

## 26. THE LAND OF DOY

*All ages*

*Content*
number, logic

*Materials*
currency templates, pencils, coloring materials

*Why?*
- Help relate number concepts to everyday life.
- Develop life skills around work and reward.

*How?*
The purpose of this activity is to link mathematics to the real world by establishing an economy and a currency in the classroom. There are two sections to this activity: The creation of a *currency*, which includes coins and/or paper money, and the establishment of an *economy* in the classroom country of your choice (ours is the Land of Doy).

*Step 1:* In the first part of the activity students, working individually or in groups, design currency to be used in their classroom country. Students need to determine denominations of the currency as well as actually to design it.

*Step 2:* A variety of related activities connected with other curriculum content areas can be tied into this activity. Students can explain (in writing or orally) the history of their currency, the meaning of the symbolism chosen for the design, and the rationale for any of the other decisions they have made, for example, size, shape, etc. Encourage students to write about the history and geographic placement of their "fantasty country." Students can write reports on their country, design flags, or make maps. The possibilities are limited only by imagination and, of course, time.

*Progression 1*

The second part of this activity involves creating a mini-economy inside the classroom. This can be done by choosing one set of bills and/or coins for the class and conducting "business" for a specified period with the new "coin of the realm."

To do this you or students need to duplicate sufficient quantities of the currency. The currency goes into circulation when you pay students for completed assignments, attendance, and other various *work* that occurs in the classroom. In addition to earning money, students can be fined for tardiness, late work, misconduct, or any violation upon which you or the students decide.

This activity works best when students are given the opportunity to generate the rules under which the economy will operate. That is, how much each *good* deed is worth, and how much each *bad* deed should cost. Students can brainstorm many ways to employ the currency such as paying for extra recess or purchasing pencils and other supplies.

*Progression 2*

Conduct a Friday "flea market" where students bring homemade and unwanted items from home and offer them for sale or barter. This is a nice way to recycle.

*Progression 3*

As you come to the end of the time allocated for your class economy (all good things must come to an end) an exciting activity that helps children dispose of their money, as well as providing practice adding and subtracting, is an auction. The items for this auction can be purchased or donated. You, or a designated student auctioneer, can hold items up for bid. You may wish to establish ground rules such as bid increases must increase by five or ten Doy dollars.

Such an auction can be quite a lot of fun. If you run out of time or energy before you run out of goods, you may wish to price items remaining for quick sale.

*Thinking It Over*

This activity can take over the classroom for weeks, even the entire year. It is worth it, especially when students share a major part of the responsibility for creating, organizing, and running the economy. There are countless subtle and often not so subtle lessons being learned here. If you and your students are patient and work together, your classroom's fantasy country and its economic system can be one of the most rewarding classroom events of the year.

STUDENT RESOURCE MATERIAL

Adams, Barbara. *The Go-Around Dollar*. Four Winds Press, 1992.
Briers, Audrey. *Money*. Franklin Watts, 1987.
Cantwell, Lois. *Money and Banking*. Franklin Watts, 1984.
Rockwell, Thomas. *How to Get Fabulously Rich*. Dell, 1991.
Schwartz, David. *How Much Is a Million?* Scholastic, 1987.
_____. *If You Made a Million*. Lothrop, Lee & Shepard, 1989.
Viorst, Judith. *Alexander Who Used to Be Rich Last Sunday*. Macmillan, 1989.

# 27. STUDENT STATISTICS

*All ages*

*Content*
probability and statistics, logic

*Materials*
pencil, paper

*Why?*
- See the use of numbers in real-world contexts.
- Gain experience collecting and interpreting data.
- Learn something about one another.

*How?*
The purpose of this activity is to give students the opportunity to work with statistics and statistical concepts.

*Step 1:* Begin by asking students to brainstorm a list of characteristics of themselves that can be numbered, quantified, or measured. Such a list might include: address, phone number, height, weight (mass), shoe size, favorite or lucky number, number of people in their families. The list is literally endless.

*Step 2:* From this initial list, students choose a set of topics to investigate. Have students generate several questions that they would like to answer about these topics. For example, if the topic is shoe size, the questions might include: "What is the most common shoe size in our class? What is the largest shoe shoe size? What is the smallest shoe size?"

. . . . . . . . . . . . . .

## 28. BURNING QUESTIONS

*All ages*

*Content*
number, probability and statistics, logic

*Materials*
paper, pencils, crayon, marking pens

*Why?*
- Provide opportunities to develop critical thinking: drawing conclusions, synthesizing information, making inferences and predictions.

*How?*
The purpose of this activity is to have students decide upon and answer "burning questions" that are of interest to them. For example: "Who is the most popular professional athlete of the students in the class?" "What animal is your favorite pet?" Students work in pairs or small groups.

The nature of the questions that students generate will, of course, vary by age and developmental level. In the younger grades students often choose questions such as: "How many brothers and sisters do you have? What is your favorite T.V. show?" As students' language abilities develop, encourage them to choose more complex questions. For example: "What percentage of students ride their bikes to school? Walk? Are driven?" As the content and the sophistication of the questions increase, students have to find and develop more sophisticated tools for gathering their data, interpreting the results, and drawing conclusions.

*Step 1:* Students must frame a question. You can help by modeling if you feel this is necessary, but the best method is for students to dive right in. Students soon discover that some question frames work better than others. For example, choosing a frame such as "What is your favorite food?" provokes so many possible responses that collection and display is very difficult. Students soon learn to limit frames: "Which of these foods is your favorite: pizza, hamburger?," etc. (supplying a specific number).

*Step 2:* Students must decide upon a way to gather the information necessary to answer their questions. How will they survey the group? Will they ask everyone? Do they need to? How will they keep from asking someone twice? Let them make these discoveries.

*Step 3:* Students organize information collected into a visual display. Keep in mind that there are many colorful and unique ways to display information beyond the common bar graph. Encourage students to be as creative and as different as possible in the display of their data. Pie charts, line graphs, displays that use the shape of the specific information being gathered are all possibilities. For example, students might use a picture of a human body to represent the different sizes of students in the class — the taller the student, the longer in length the representation.

*Step 4:* Students interpret and draw conclusions about the information that they have collected and displayed. Encourage the entire class to analyze the information that groups or paired students present. Have them form statements that capture the information in the display as well as frame additional questions appropriate to the information. Does everyone agree with the statements and conclusions? It helps if you model good questions at this point in the activity:

*About a pets display*
For our class, which is the favorite pet?
Does this mean all children love cats best?
What other kinds of pets would a questionnaire include if we lived in a different country or culture?

If my favorite pet is an elephant, where does my answer go?

*About a television display*

What does this graph tell us about our favorite night for television?

What kinds of shows does our class like best?

Do you think the results would be the same for parents? The goal is to have students develop the ability to form their own questions and conclusions. A question such as: "Look at this graph and ask a question based on the information we've gathered," helps facilitate this process.

These displays, hung about the room, make for exciting, colorful testimony to student interest and learning. Discussion often continues sparking new questions to be explored.

## Progression

The activity can be extended by having students create written statements synthesizing the information depicted in their samples. With older students, their data can lead to discussion and lessons about percentages, consumerism, likes and differences, etc.

The activity also lends itself to integration with computer activities. Many commercially available software packages provide tools for the display of information in a variety of formats.

## Thinking It Over

*Burning Questions* is a useful, fun, and challenging language activity. The activity gives children experience in forming questions, gathering information (statistics), and subsequently interpreting and drawing inferences from the data. These skills are essential for understanding our modern world.

STUDENT RESOURCE MATERIAL

Arnold, Caroline. *Charts and Graphs: Facts and Activities*. Franklin Watts, 1984.

Dunworth, John, and Thomas Drysdale. *Millions of People*. Holt, Rinehart & Winston, 1971.

James, Elizabeth, and Carol Barkin. *What Do You Mean By ''Average''? Means, Medians and Modes*. Lothrop, Lee and Shepard, 1978.

# 29. MYSTERY GRAPHS: OPEN BAR

*All ages*

*Content*
problem solving, logic, probability and statistics

*Materials*
copies of graph sheets (see end of this activity), paper, pencils

*Why?*
• Gain experience working with concepts used in graphing.
• Encourage divergent and open-ended thinking.

*How?*
The purpose of this activity is to help students understand the many ways that comparative information can be represented through graphing.

Be sure that the students have worked with bar graphs. They should already have had experience gathering information and transferring it to graphs and diagrams of their own creation (see *Burning Questions* and *Student Statistics*).

Have students work in groups to develop interpretations and explanations for the first three graphs provided. For example, in Mystery Bar Graph 1 there are three values represented. They could represent: the allowances of siblings in a family, ages of pets, television hours watched, numbers of shoes in a closet, or sports scores. What else could they represent?

Have groups come up with one written response to the graph. Encourage diversity. Encourage creative explanations. For a challenge have groups use the same graph and develop several different explanations. Have each group present their work to the whole class.

*Progression 1*

After the class is comfortable with the process move to Mystery Bar Graphs 4 and 5. Students are to fill in the values on the vertical axis of the graph as well as come up with a subject. These graphs encourage an even greater range of possibilities.

*Progression 2*

Bar graphs are only one way to display information. Have them experiment with other forms. Introduce and discuss pie charts and line graphs. Once introduced to these various modes, students can transfer the information on their bar graphs to these other formats.

Encourage creativity with materials as well. Have groups create graphs with colored paper and marking pens. Display one graph and have all groups interpret the same graph in their groups. Share the results. Post the one graph on the bulletin board and surround it with different interpretations. It's colorful, fun, and yet another reason to read!

*Progression 3*

Don't stop here. Have groups design their own Mystery Graphs for others to interpret.

*Thinking It Over*

The ability to generate, analyze and interpret information is a critical ability in our society. Mystery graphs provide students with an opportunity to learn the basic concepts associated with graphing and information display while at the same time requiring them to link that knowledge to their lives. When students use the language and concepts associated with investigative graphing they are developing real life skills.

STUDENT RESOURCE MATERIAL

Petty, Kate. *Numbers.* Franklin Watts, 1986.

**Note:** *For student use, enlarge each activity sheet by 130-150%.*

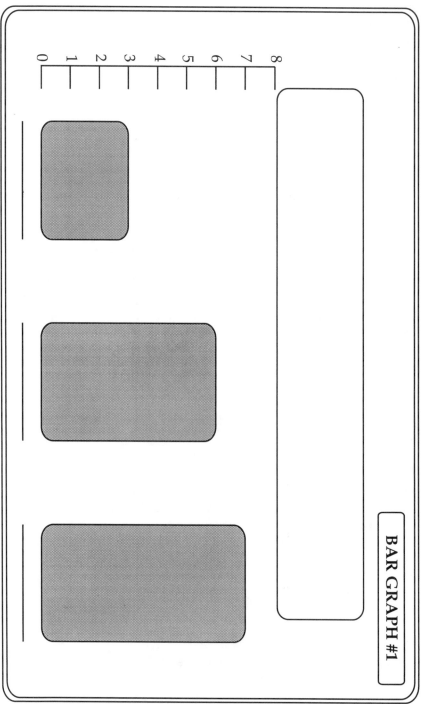

BAR GRAPH #1

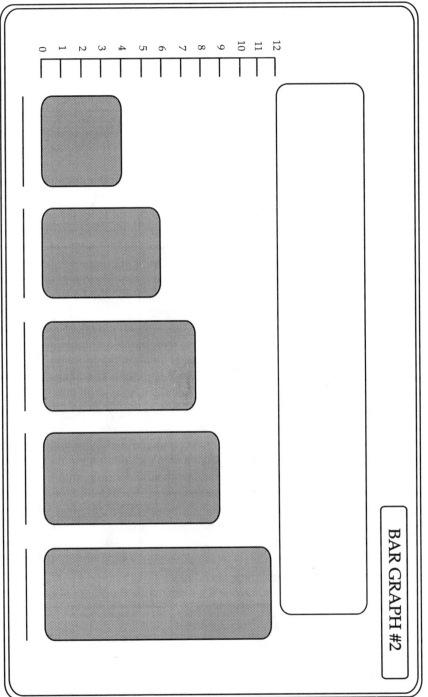

BAR GRAPH #2

123

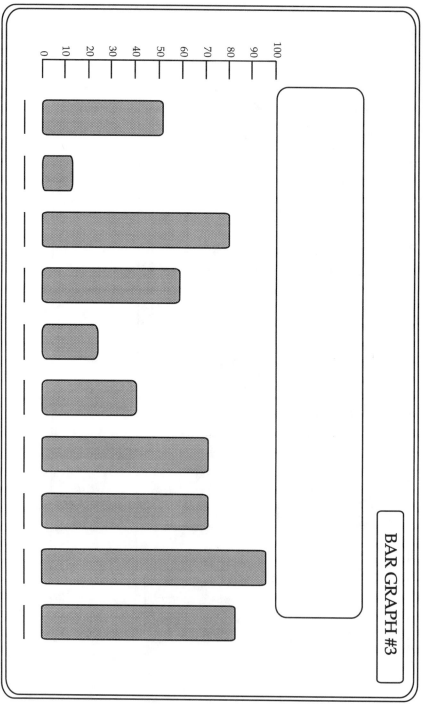

BAR GRAPH #3

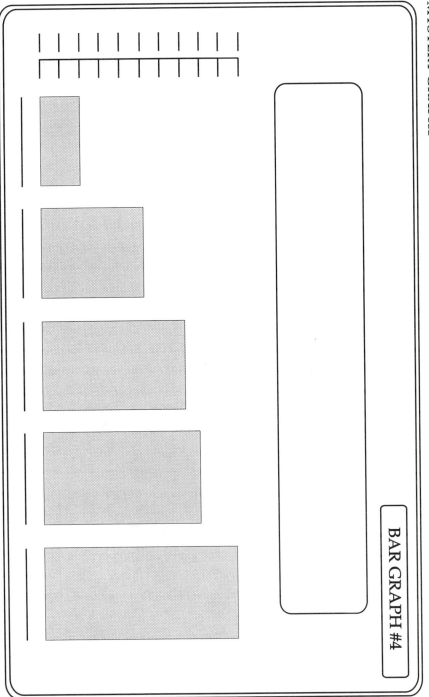

BAR GRAPH #4

MYSTERY GRAPHS

BAR GRAPH #5

126

. . . . . . . . . . . . .

## 30. MYSTERY GRAPHS: NATURAL PHENOMENA

*Ages 9–12*

*Content*
problem solving, logic, probability and statistics

*Materials*
Copies of graph sheets (see end of this activity), paper, pencil, various reference materials

*Why?*
- Encourage problem solving, creative solutions, and divergent thinking.
- Gain experience with the purpose and process of graphing.

*How?*
The purpose of this activity is to help students understand the many ways that information about the natural world can be represented through graphing.

*Step 1:* Start with Graph A. Introduce it to students telling them that the statistics in this graph are taken from a group or set of something in their world (in this case the four oceans).

As a class brainstorm ideas of what might be represented by this graph. All answers are acceptable when creating this first list. Focus on what you and your students know. Ask the students what information is available from the graph as it is presented: "Whatever these are, there are four of them; they can be measured in square kilometres or miles; what could be over 100 million square kilometres in area?"

*Step 2:* Look at your brainstorm list. Decide if there are any ideas on the list that might not fit. Be sure to ask students to give their reasons for wishing to strike an item.

*Step 3:* In groups, have students work together to decide on one thing that may be represented by this graph. They should write all their arguments for why they feel their solution makes sense.

It is very important to stress with students that while having the "right" answer might be fun, it is really not of great importance to this activity. What you are looking for is logical, well defined arguments. Logical arguments are the goal of this work.

Once your students have made their final arguments and you have revealed the "answer" to the mystery graph, extend the discussion on the phenomenon in question. For example, with oceans: "Which ocean is closest to where we live? What things live in the oceans? How are people dependent upon the earth's oceans? What else do you know about oceans?" For example, with planets (graph E): "What are some other ways we measure the planets? How did they get their names? How were they discovered? What are some of the ways people relate to or use planets today?"

Each of the graphs can be used as starting points for further investigation of natural phenomena.

*Step 4:* When students have worked through several of the graphs and understand the process, have them work in their groups to develop their own displays for other groups to work with. They can use encyclopedias, almanacs, record books, or other reference materials.

*Thinking It Over*

This activity presents a way to introduce factual information to the class. In addition, when groups are developing their own graphs, they will need to use various reference materials to make their work accurate. They will be learning about many general ideas of interest, practising their reference skills and having fun.

STUDENT RESOURCE MATERIAL

Parker, Tom. *In One Day.* Houghton Mifflin, 1984.

**Note:** *For student use, enlarge each activity sheet by 130-150%.*

# NATURAL PHENOMENA FACTS

*Display A: Oceans of the Earth (km²/square miles)*

| | |
|---|---|
| Pacific | 166 884 000 / 63 800 000 |
| Atlantic | 86 892 000 / 31 530 000 |
| Indian | 73 711 000 / 28 356 000 |
| Arctic | 13 275 000 /  3 662 000 |

*Display B: Speed of various land animals from slow to fast (km/h/mph)*

| | | | | | |
|---|---|---|---|---|---|
| Tortoise | 16/.10 | Snake | 3/2 | Human | 32/20 |
| Elephant | 40/25 | House Cat | 48/30 | Fox | 64/40 |
| Ostrich | 64/40 | Greyhound | 64/40 | Racehorse | 72/45 |
| Jackrabbit | 72/45 | Gazelle | 80/50 | Cheetah | 110/70 |

*Display C: Whales in alphabetical order (metres/feet)*

| | | | | | |
|---|---|---|---|---|---|
| Beluga | 5/15 | Blue | 30/100 | Bowhead | 18/60 |
| Fin | 24/80 | Grey | 15/50 | Humpback | 15/50 |
| Minke | 9/30 | Narwhal | 5/15 | Orca | 9/30 |
| Pilot | 8.5/28 | Right | 18/60 | Sperm | 18/60 |

*Display D: Continents of Earth (km²/square miles)*

| | |
|---|---|
| Asia | 44 045 000 / 17 006 000 |
| Africa | 30 261 000 / 11 684 000 |
| North America | 24 237 000 /  9 358 000 |
| South America | 17 806 000 /  6 875 000 |
| Antarctica | 14 000 000 /  5 400 000 |
| Europe | 10 521 000 /  4 062 000 |
| Australia | 7 682 000 /  2 966 000 |

*Display E: Planets in our solar system in order from the sun.*
*Circumference at their equators (kilometres/miles).*

| | | | |
|---|---|---|---|
| Mercury | 15 316 /  9 517 | Saturn | 378 483 / 235 179 |
| Venus | 38 006 / 23 615 | Uranus | 160 510 /  99 735 |
| Earth | 40 053 / 25 571 | Neptune | 155 430 /  96 712 |
| Mars | 21 339 / 13 260 | Pluto | 7 222 /  4 490 |
| Jupiter | 448 970 / 278 976 | | |

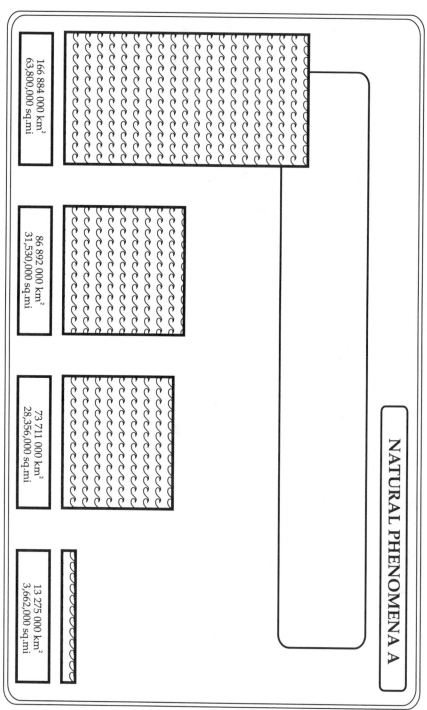

166 884 000 km²
63,800,000 sq.mi.

86 892 000 km²
31,530,000 sq.mi.

73 711 000 km²
28,356,000 sq.mi.

13 275 000 km²
3,662,000 sq.mi.

NATURAL PHENOMENA A

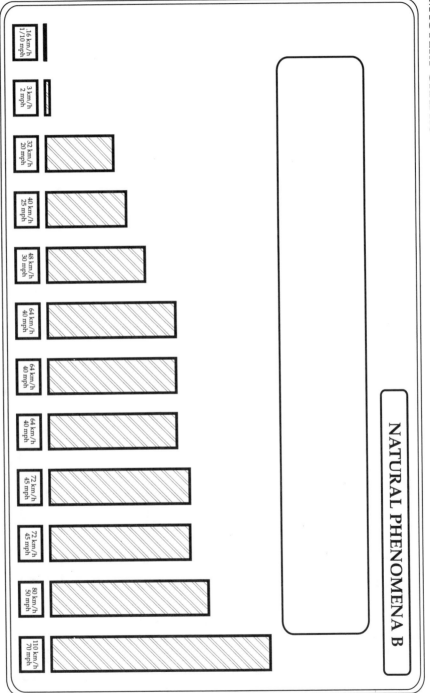

NATURAL PHENOMENA B

.16 km/h
1/10 mph

3 km/h
2 mph

32 km/h
20 mph

40 km/h
25 mph

48 km/h
30 mph

64 km/h
40 mph

64 km/h
40 mph

64 km/h
40 mph

72 km/h
45 mph

72 km/h
45 mph

80 km/h
50 mph

110 km/h
70 mph

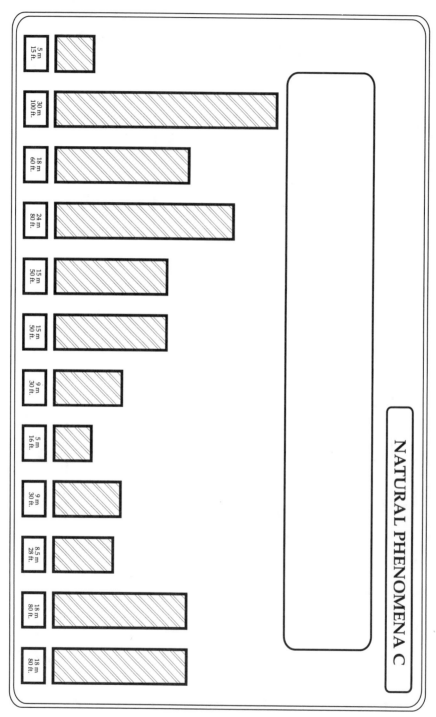

NATURAL PHENOMENA C

| | |
|---|---|
| 5 m<br>15 ft. | |
| 30 m<br>100 ft. | |
| 18 m<br>60 ft. | |
| 24 m<br>80 ft. | |
| 15 m<br>50 ft. | |
| 15 m<br>50 ft. | |
| 9 m<br>30 ft. | |
| 5 m<br>16 ft. | |
| 9 m<br>30 ft. | |
| 8.5 m<br>28 ft. | |
| 18 m<br>80 ft. | |
| 18 m<br>80 ft. | |

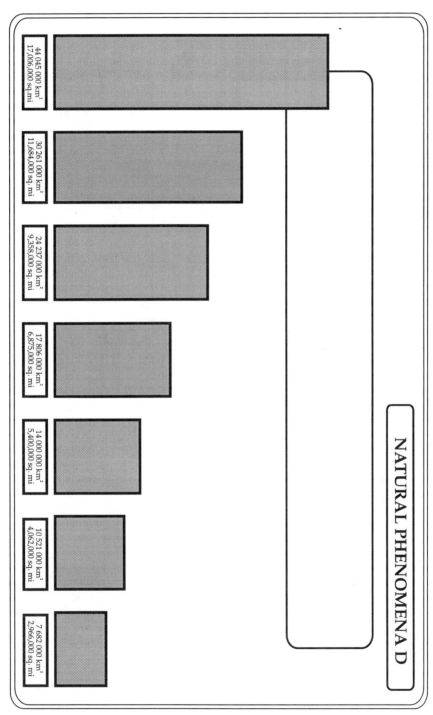

NATURAL PHENOMENA D

44 045 000 km²
17,006,000 sq. mi.

30 261 000 km²
11,684,000 sq. mi.

24 237 000 km²
9,358,000 sq. mi.

17 806 000 km²
6,875,000 sq. mi.

14 000 000 km²
5,400,000 sq. mi.

10 521 000 km²
4,062,000 sq. mi.

7 682 000 km²
2,966,000 sq. mi.

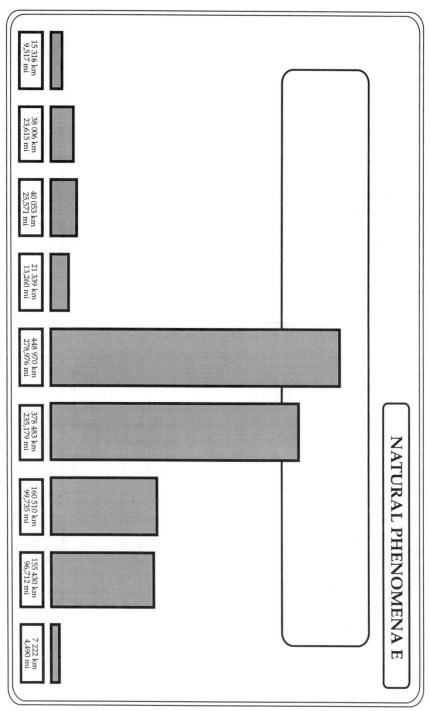

NATURAL PHENOMENA E

15 316 km
9,517 mi

38 006 km
23,615 mi

40 053 km
25,571 mi

21 339 km
13,260 mi

448 970 km
278,976 mi

378 483 km
235,179 mi

160 510 km
99,735 mi

155 430 km
96,712 mi

7 222 km
4,490 mi

## 31. MYSTERY GRAPHS: THE PLOT THICKENS

*Ages 8–12*

*Content*
problem solving, logic, probability and statistics

*Materials*
copies of graph sheets (see end of this activity), paper, pencils

*Why?*
• Gain experience plotting an event over time.
• Develop sense of temporal sequence.

*How?*
The purpose of this activity is to help students understand the ways that temporal events can be represented through line graphing.

*Step 1:* Students work to interpret familiar events as they take place over time based on a line graph. For example, in Mystery Line Graph 1, *Students in the Room*, the graph shows how many students are in the room from the beginning of the day through to the end of the school day.

Have groups work together to explain what is happening to cause the population to fluctuate. Have them write a description or story of the day's events that chronicles the reasons for the changes in classroom population.

Be sure to take time to brainstorm what events could be taking place to cause students to enter and leave the room at different times. Encourage a wide range of ideas. Have students work in groups to develop their own stories interpreting the graph. An important topic during discussion is how the groups kept track of the times relative to the event as they worked through the day.

Repeat the same procedure for the remaining graphs.

*Step 2:* Once the classroom has explored these graphs, have them create their own. Think of other situations that change over time: people at a bus stop, flowers in the garden, cars on the street, etc.

## Thinking It Over

These activities encourage problem solving and creative solutions. Students enjoy the open-endedness of applying their own experiences in situations where there are really no right or wrong answers. This encourages divergent thinking.

STUDENT RESOURCE MATERIAL

Allen, Pamela. *Mr. Archimedes' Bath*. Harper Collins, 1991.

Stwertka, Eve and Albert Stwertka. *Make it Graphic! Drawing Graphs for Science and Social Studies Projects*. Julian Messner, 1985.

**Note:** *For student use, enlarge each activity sheet by 130-150%.*

THE PLOT THICKENS #1

STUDENTS IN THE ROOM

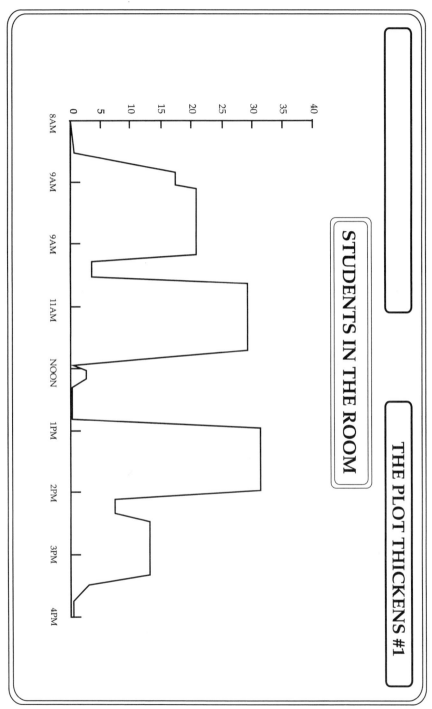

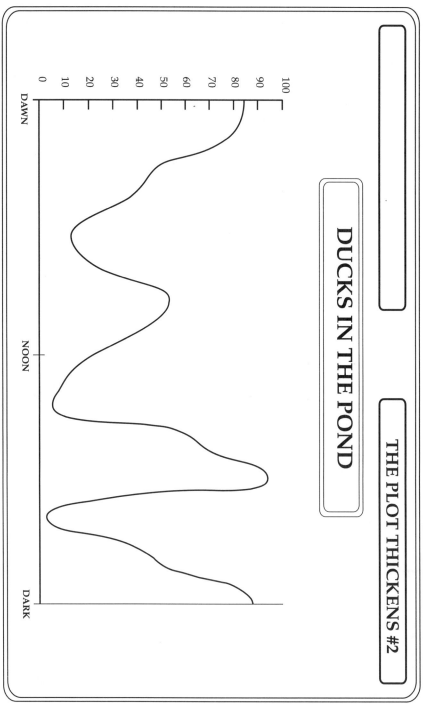

THE PLOT THICKENS #2

DUCKS IN THE POND

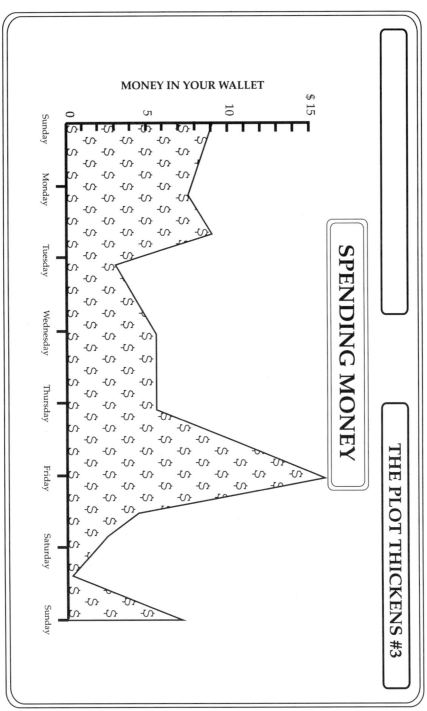

MONEY IN YOUR WALLET

SPENDING MONEY

THE PLOT THICKENS #3

139

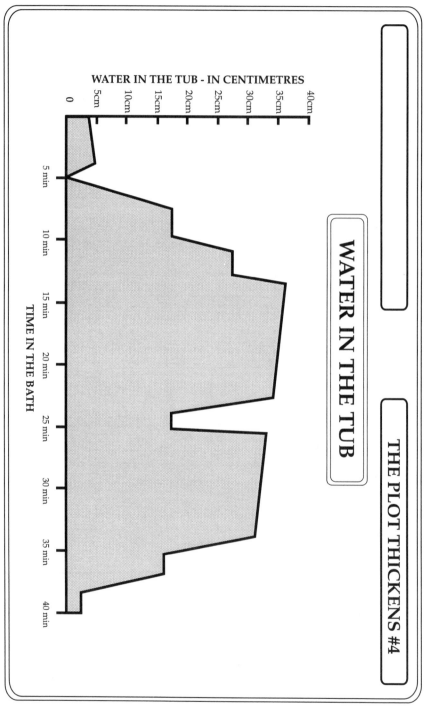

MYSTERY GRAPHS

THE PLOT THICKENS #4

WATER IN THE TUB

WATER IN THE TUB - IN CENTIMETRES

0 · 5cm · 10cm · 15cm · 20cm · 25cm · 30cm · 35cm · 40cm

5 min · 10 min · 15 min · 20 min · 25 min · 30 min · 35 min · 40 min

TIME IN THE BATH

140

AN EXCHANGE OF GIFTS:
A STORYTELLER'S HANDBOOK
Marion V. Ralston

*Imaginative activities to enhance language programs*
*by promoting classroom storytelling.*

THE WORD WALL: TEACHING VOCABULARY
THROUGH IMMERSION
Joseph Green

*Using mural dictionaries — word lsits on walls — to strengthen*
*children's reading, speaking and writing skills.*

INFOTEXT: READING AND LEARNING
Karen M. Feathers

*Classroom-tested techniques for helping students overcome*
*the reading problems presented by informational texts.*

WRITING IN THE MIDDLE YEARS
Marion Crowhurst

*Suggestions for organizing a writing workshop approach*
*in the classroom.*

AND THEN THERE WERE TWO: CHILDREN
AND SECOND LANGUAGE LEARNING
Terry Piper

*A practical resource answering many*
*of the most commonly asked questions of*
*English as a second language teachers.*

IN ROLE: TEACHING AND LEARNING
DRAMATICALLY
Patrick Verriour

*One of the English-speaking world's*
*leading drama educators demonstrates,*
*with practical examples, how easily*
*drama can be used for integrating*
*learning across the curriculum.*